LUBRICANTS

LUBRICANTS
INTRODUCTION TO PROPERTIES AND PERFORMANCE

Marika Torbacke
Statoil Lubricants, Sweden

Åsa Kassman Rudolphi
Uppsala University, Sweden

Elisabet Kassfeldt
Luleå University of Technology, Sweden

This edition first published 2014
© 2014 John Wiley & Sons Ltd

Registered office
John Wiley & Sons Ltd, The Atrium, Southern Gate, Chichester, West Sussex, PO19 8SQ, United Kingdom

For details of our global editorial offices, for customer services and for information about how to apply for permission to reuse the copyright material in this book please see our website at www.wiley.com.

The right of the author to be identified as the author of this work has been asserted in accordance with the Copyright, Designs and Patents Act 1988.

All rights reserved. No part of this publication may be reproduced, stored in a retrieval system, or transmitted, in any form or by any means, electronic, mechanical, photocopying, recording or otherwise, except as permitted by the UK Copyright, Designs and Patents Act 1988, without the prior permission of the publisher.

Wiley also publishes its books in a variety of electronic formats. Some content that appears in print may not be available in electronic books.

Designations used by companies to distinguish their products are often claimed as trademarks. All brand names and product names used in this book are trade names, service marks, trademarks or registered trademarks of their respective owners. The publisher is not associated with any product or vendor mentioned in this book.

Limit of Liability/Disclaimer of Warranty: While the publisher and author have used their best efforts in preparing this book, they make no representations or warranties with respect to the accuracy or completeness of the contents of this book and specifically disclaim any implied warranties of merchantability or fitness for a particular purpose. It is sold on the understanding that the publisher is not engaged in rendering professional services and neither the publisher nor the author shall be liable for damages arising herefrom. If professional advice or other expert assistance is required, the services of a competent professional should be sought.

Library of Congress Cataloging-in-Publication Data applied for.

ISBN 9781118799741

Set in 10/12pt Times by Aptara Inc., New Delhi, India
Printed and bound in Malaysia by Vivar Printing Sdn Bhd

1   2014

Contents

Preface	xi
List of Symbols	xiii
List of Tables	xvii

Part One   LUBRICANT PROPERTIES

1	**Introduction to Tribology**		**3**
1.1	Tribological Contacts		5
	1.1.1	*Macroscale Contacts*	6
	1.1.2	*Microscale Contacts*	8
1.2	Friction		8
	1.2.1	*The Coefficient of Friction*	8
	1.2.2	*Lubrication Regimes*	10
1.3	Wear		12
	1.3.1	*Wear Rate*	13
1.4	Lubrication of the Tribological System		14
	1.4.1	*The Purposes of Lubricants*	14
	1.4.2	*Reducing Friction and Protecting against Wear*	15
	1.4.3	*Semi-Solid Lubricants*	16
	1.4.4	*Solid Lubricants and Dry Lubricants*	16
	References		17
2	**Lubricant Properties**		**19**
2.1	Performance Properties		20
	2.1.1	*Viscosity*	20
	2.1.2	*Low and High Temperature Properties of Lubricants*	27
	2.1.3	*Air and Water Entrainment Properties*	29
	2.1.4	*Thermal Properties*	32
2.2	Long Life Properties		33
	2.2.1	*Total Acid Number (TAN)*	34
	2.2.2	*Total Base Number (TBN)*	35
	2.2.3	*Oxidation Stability*	35

	2.2.4	Hydrolytic Stability	37
	2.2.5	Corrosion Inhibition Properties	37
2.3	Environmental Properties		40
	2.3.1	Environmentally Adapted Lubricants	40
	2.3.2	Market Products with a Reduced Environmental Impact	41
2.4	Summary of Analyses		42
	References		44
3	**Base Fluids**		**45**
3.1	General Hydrocarbon Chemistry		45
3.2	Base Fluid Categorization		48
3.3	The Refining Process of Crude Oils		50
	3.3.1	The Refining Process	51
	3.3.2	Influence of the Refining Process on the Oil Properties	52
3.4	Base Fluids Originating from Crude Oil		53
	3.4.1	Paraffinic Base Oils	53
	3.4.2	Naphthenic Base Oils	53
	3.4.3	White Oils	54
	3.4.4	Very High Viscosity Index Base Oils	54
	3.4.5	Polyalphaolefins	54
	3.4.6	Gas-to-Liquid Base Fluids	55
	3.4.7	Re-Refined Base Oils	56
3.5	Base Fluids Originating from Renewable Raw Materials		56
	3.5.1	Vegetable Oils (Natural Esters)	57
	3.5.2	Synthetic Esters	57
3.6	Nonconventional Synthetic Base Fluids		59
3.7	Properties of Base Fluids		59
	References		61
4	**Additives**		**63**
4.1	Fundamental Concepts and Processes		63
	4.1.1	Atoms and Reactions	63
	4.1.2	Intermolecular Forces	64
	4.1.3	Chemical Potential	66
	4.1.4	Surfaces	66
	4.1.5	Mass Transfer	67
	4.1.6	Adsorption	68
	4.1.7	Chemical Characteristics of Surface Active Additives	70
4.2	Additive Exploration		71
4.3	Surface Active Adsorbing Additives		73
	4.3.1	Corrosion Inhibitors	73
	4.3.2	Friction Modifiers	75
	4.3.3	Antiwear Additives	75
	4.3.4	Extreme Pressure Additives	76
	4.3.5	Activation of Antiwear and Extreme Pressure Additives	77
	4.3.6	Competition for Surface Sites by Surface Active Additives	78

4.4	Interfacial Surface Active Additives		79
	4.4.1	Defoamers	79
	4.4.2	Emulsifiers and Demulsifiers	80
4.5	Physically Bulk Active Additives		81
	4.5.1	Viscosity Modifiers	81
	4.5.2	Pour Point Depressants	82
	4.5.3	Dispersants	84
4.6	Chemically Bulk Active Additives		85
	4.6.1	Detergents	85
	4.6.2	Antioxidants	87
4.7	Additive Summary		88
	References		89

Part Two LUBRICANT PERFORMANCE

5	**Formulating Lubricants**		**93**
5.1	General Aspects of Development		93
	5.1.1	Formulations	93
	5.1.2	Development Work	96
	5.1.3	Material Compatibility	96
	5.1.4	Miscibility	97
	5.1.5	Interactions in a Lubricated Contact	97
5.2	Quality of the Lubricated Tribological Contact		98
	5.2.1	Lubricant Film Regime	99
	5.2.2	Maintaining a High Quality Contact	101
5.3	Hydraulics		101
	5.3.1	Description of a Hydraulic System	101
	5.3.2	Formulating Hydraulic Oils	102
5.4	Gears		104
	5.4.1	Description of Gears	104
	5.4.2	Formulating Gear Oils	105
5.5	Combustion Engines		107
	5.5.1	Description of Combustion Engines	107
	5.5.2	Formulating Combustion Engine Oils	108
	References		110

6	**Tribological Test Methods**		**113**
6.1	Field, Bench and Component Tests		113
6.2	Model Tests		115
	6.2.1	Strategy for Selecting and Planning a Model Test	115
6.3	Lubricant Film Thickness Measurements		117
	6.3.1	Electrical Methods	117
	6.3.2	Optical Interferometry Method	118
6.4	Tribological Evaluation in Mixed and Boundary Lubrication		121
	6.4.1	The Pin-on-Disc Tribotest	121

		6.4.2	The Reciprocating Tribotest	123
		6.4.3	The Twin Disc Tribotest	124
		6.4.4	The Rotary Tribotest	128
	6.5	Selection of Model Tests to Simulate Real Contacts		128
		6.5.1	Hydraulics	129
		6.5.2	Gears	129
		6.5.3	Combustion Engines	131
	6.6	Summary of Tribotest Methods		131
		References		132
7	**Lubricant Characterization**			**133**
	7.1	General Characterization Concepts		133
		7.1.1	Planning	133
		7.1.2	Basic Mixing Theory	134
		7.1.3	Sampling	135
		7.1.4	Diluting the Sample	136
		7.1.5	Collecting Analysis Data	137
		7.1.6	Calculations and Evaluation	138
	7.2	Condition Analyses of Lubricants		138
	7.3	Nonused Oil Characterization		140
		7.3.1	Development	140
		7.3.2	Production	141
		7.3.3	Application Examples	142
	7.4	Used Oil Characterization		142
		7.4.1	Selection of Analyses	143
		7.4.2	Analysis Examples of Selected Applications	144
	7.5	Summary of Used Oil Analyses		146
		References		148
8	**Surface Characterization**			**149**
	8.1	Surface Characterization of Real Components		151
		8.1.1	Examination of Nonused Surfaces	151
		8.1.2	Examination of Used Surfaces	151
		8.1.3	Characteristics of Application Examples	152
	8.2	Microscopy Techniques		153
		8.2.1	Visual Inspection	153
		8.2.2	Light Optical Microscopy (LOM)	154
		8.2.3	Optical Interference Microscopy	154
		8.2.4	Atomic Force Microscopy (AFM)	154
		8.2.5	Scanning Electron Microscopy (SEM)	155
		8.2.6	Focused Ion Beam (FIB)	158
		8.2.7	Transmission Electron Microscopy (TEM)	159
	8.3	Surface Measurement		159
		8.3.1	Statistical Surface Parameters	161
		8.3.2	Contacting Stylus Profiler	162
		8.3.3	Microscopy Techniques	163

8.4	Hardness Measurement		163
	8.4.1	*Macro and Micro Hardness*	163
	8.4.2	*Nanoindentation*	163
8.5	Surface Analysis Techniques		163
	8.5.1	*Selected Methods*	164
	8.5.2	*Analysis Performance Parameters and Terminology*	165
	8.5.3	*Depth Profiling and Chemical Mapping*	167
	8.5.4	*Energy Dispersive X-Ray Spectroscopy (EDS)*	169
	8.5.5	*Auger Electron Spectroscopy (AES)*	170
	8.5.6	*X-Ray Photoelectron Spectroscopy (XPS)*	173
	8.5.7	*Secondary Ion Mass Spectroscopy (SIMS)*	176
	8.5.8	*Fourier Transform Infrared Spectroscopy*	178
8.6	Summary of Surface Characterization Methods		179
	8.6.1	*Microscopy and Surface Measurement*	179
	8.6.2	*Surface Analysis*	179
	References		182

Index   **185**

Preface

How to lubricate for optimal and lasting function is a key challenge. Today there is a trend towards higher power density applications, which also have to meet increasing environmental demands. This requires an understanding of both the applications and the lubricants. *Lubricants: Introduction to Properties and Performance* is written to fulfil the gap between the knowledge of applications and that of lubricants. It is divided into two parts. The first part is theoretical and serves as the basis for understanding. It describes lubricant properties as well as base fluids and additives. The second part is hands-on and introduces the reader to the actual formulations and the evaluation of their performance.

The first part comprises four chapters:

- Introduction to Tribology
- Lubricant Properties
- Base Fluids
- Additives

The chapter *Introduction to Tribology* introduces the reader to the tribological contact, friction, wear and lubrication. *Lubricant Properties* introduces the basic concepts regarding properties and the most commonly made analyses on lubricants. *Base Fluids* explores the origin of different base fluids and their properties. In the same manner the chapter *Additives* gives the reader a flavour of the broad spectrum of common additives used in lubricants.

The second part comprises four chapters:

- Formulating Lubricants
- Tribological Test Methods
- Lubricant Characterization
- Surface Characterization

Formulating Lubricants describes different applications and the corresponding lubricant formulations necessary to fulfil their requirements. In *Tribological Test Methods* the level of testing is discussed followed by a focus on model testing in a laboratory environment. The final chapters, *Oil Characterization* and *Surface Characterization*, give the reader an introduction to different methods of characterizing lubricants and surfaces respectively.

Writing *Lubricants: Introduction to Properties and Performance* has been a joint effort of Statoil Lubricants, Luleå University of Technology and Uppsala University. The aim was to write the tribological parts for a person with a chemical engineering background and the lubricant parts for a person with a mechanical engineering background. We have tried to write the book in an easy-to-read style. The theoretical discussion is kept to a minimum level of chemistry and physics, only introduced when they are required for the reader to understand the related principles or performance outputs.

Finally, writing the book has been very rewarding work for us, but this work has reached a higher level with the knowledge from others. We would therefore like to thank Prof. Sture Hogmark (Uppsala University), Prof. Erik Höglund (Luleå University of Technology), Prof. Thomas Norrby (Statoil Lubricants) and Pär Nyman (Statoil Lubricants) for reviewing and valuable commenting on the material at an early stage of the work. Several persons have been very helpful during the writing by contributing their knowledge in special areas. Among these are colleagues at R&D (Statoil Lubricants), as well as colleagues at the tribomaterials group (Uppsala University) and at the machine elements group (Luleå University of Technology). Thanks, to all those who have helped us.

We hope that the book is both educational and can serve as a handbook for ready reference. Enjoy your reading of *Lubricants: Introduction to Properties and Performance*!

Marika Torbacke, Åsa Kassman Rudolphi and Elisabet Kassfeldt
Sweden, November 2013

List of Symbols

The list of symbols includes a description and the section in which a symbol is introduced or explained.

Symbol	Description	Unit	Section
A	contact area	m²	1.2.1, 2.1.4
AES	Auger electron spectroscopy	–	8.5.5
AFM	atomic force microscopy	–	8.2.4
ATF	automatic transmission fluid	–	4.3.2, 4.3.3, 4.5.2, 4.5.3
AW	antiwear additive	–	4.3.3, 4.3.5, 4.3.6
c	concentration	mole/m³	4.1.5
c_{ads}	ratio of concentrations in solution and at surface	–	4.1.6
$c_{solution}$	concentration in the solution	mole/m³	4.1.6
$c_{surface}$	concentration at the surface	mole/m³	4.1.6
const	constant used in Arrhenius equation	–	2.2.3
C_nH_{2n+2}	hydrocarbon chain with numbers indicated	–	3.1
C_xH_y	hydrocarbon chain	–	2.2.3, 3.1
C_p	specific heat capacity	J/kg, K	2.1.4
CCS	cold cranking simulator	–	2.4, 5.5.2
D	diffusivity, diffusion coefficient	m²/s	4.1.5
E_A	activation energy	J/mole	2.2.3
EAL	environmentally adapted lubricant	–	2.3.1, 3.2, 3.5
EDS	energy dispersive X-ray spectroscopy	–	8.5.4
EHD	elastohydrodynamic lubrication	–	1.2.2
EP	extreme pressure additive	–	4.3.4, 4.3.5, 4.3.6
ESCA	electron spectroscopy for chemical analysis	–	8.5.6
F	friction force	N	1.2.1
F_{load}	externally applied load	N	1.2.1
FIB	focused ion beam	–	8.2.6

Symbol	Description	Unit	Section
FM	friction modifier additive	–	4.3.2, 4.3.6
FTIR	Fourier transform infrared	–	7.2, 8.5.8
g	gravitational constant	m/s^2	1.2.1
G	heat transfer	W	2.1.4
Group I	group I base oils	–	3.2, 3.7
Group II	group II base oils	–	3.2, 3.7
Group III	group III base oils	–	3.2, 3.7
Group IV	group IV base oils	–	3.2, 3.7
Group V	group V base oils	–	3.2, 3.7
GTL	gas-to-liquid base oil	–	3.2, 3.4.6
h	lubricant film thickness	m	1.2.2, 5.2.1
HTHS	high temperature high shear	–	2.4, 5.5.2
HV	hardness Vickers	N/m^2	8.4.1
HK	hardness Knoop	N/m^2	8.4.1
i	position in surface roughness calculations	–	8.3.1
ICP	inductively coupled plasma	–	7.2
IR	infrared	–	7.2, 8.5.8
IRS	infrared spectroscopy	–	8.1.2
k	rate constant of a chemical reaction	–	2.2.3
k_a	rate constant for adsorption	–	4.1.6
k_d	rate constant for desorption	–	4.1.6
k_w	specific wear number	–	1.3.1
K	ratio of rate constants for adsorption and desorption	–	4.1.6
KOH	potassium hydroxide	–	2.2.1, 2.2.2
KV_{40}	kinematic viscosity at 40°C	m^2/s	2.1.1.2, 2.4
KV_{100}	kinematic viscosity at 100°C	m^2/s	2.1.1.2, 2.4
LOM	light optical microscopy	–	8.2.2
m	mass	kg	1.2.1, 2.1.4
MRV	mini-rotary viscometer	–	2.4, 5.5.2
n	number of positions in surface roughness calculations	–	8.3.1
N	normal force	N	1.2.1
J	mass transfer rate	mole/m^2, s	4.1.5
NO$_x$	nitrogen oxides	–	5.5.1
OEM	original equipment manufacturer	–	3.2, 5.5.2
•OH	hydroxyl radical	–	2.2.3
p	pressure	Pa	2.1.1.4
ΔP	pressure difference	Pa	2.1.3.1
PAO	polyalphaolefin base oil	–	3.2, 3.4.5

Symbol	Description	Unit	Section
PPD	pour point depressant additive	–	4.5.2
PTFE	polytetrafluoroethylene, i.e. Teflon	–	1.4.4
Q	heat energy	J	2.1.4
r	droplet radius	m	2.1.3.1
R	gas constant	J/mole, K	2.2.3
R_a	average surface roughness	m	8.3.1
R_q	root mean square roughness	m	8.3.1
R_{qA}	roughness of surface A	m	1.2.2
R_{qB}	roughness of surface B	m	1.2.2
R	hydrocarbon chain lacking the end group	–	2.2.3
R•	radical	–	2.2.3, 4.6.2
R_1	hydrocarbon chain with index 1	–	2.2.3
R_2	hydrocarbon chain with index 2	–	2.2.3
RCOO$^-$	anion of the fatty acid	–	2.2.5.3
RCOOH	fatty acid with a hydrocarbon chain R	–	2.2.5.3, 3.5.2
R_1COOH	fatty acid with a hydrocarbon chain R_1	–	2.2.4, 3.5.2
R_2COOH	fatty acid with a hydrocarbon chain R_2	–	2.2.3
R_1COOR_2	ester with hydrocarbon chains R_1 and R_2	–	2.2.4, 3.5.2
RH	hydrocarbon chain with a hydrogen end group	–	2.2.3, 4.6.2
R_2OH	alcohol with a hydrocarbon chain R_2	–	2.2.4
RO•	radical	–	2.2.3, 4.6.2
ROO•	hydroperoxide radical	–	2.2.3, 4.6.2
ROOH	carboxylic acid	–	4.6.2
s	sliding distance	m	1.3.1
SEM	scanning electron microscopy	–	8.2.5
SIMS	secondary ion mass spectroscopy	–	8.5.7
T	temperature	K or °C	2.1.1.3, 2.1.4, 2.2.3
ΔT	temperature difference	K or °C	2.1.4
TAN	total acid number	kg KOH/kg oil	2.2.1, 2.2.3
TBN	total base number	kg KOH/kg oil	2.2.2, 4.6.1, 5.5.2
TEM	transmission electron microscopy	–	8.2.7
ToF-SIMS	time-of-flight secondary ion mass spectroscopy	–	8.5.7
t_{rc}	time for coalescence	s	2.1.3.3
u	velocity of moving body	m/s	2.1.1.2
u_{bubble}	velocity of bubble rise in lubricants	m/s	2.1.3.1
UHV	ultra-high vacuum	–	8.5.2, 8.6.2
v_a	velocity for adsorption	m/s	4.1.6
v_d	velocity for desorption	m/s	4.1.6

Symbol	Description	Unit	Section
V	wear volume	m^3	1.3.1
VI	viscosity index	–	2.1.1.3, 3.1, 4.5.1
VG	viscosity grade	–	5.3.2, 5.4.2, 5.5.2
VHVI	very high viscosity index base oil	–	3.4.4
VM	viscosity modifier additive	–	4.5.1
VSI	vertical scanning interferometry	–	8.2.3
x	distance	m	8.3.1
XPS	X-ray photoelectron spectroscopy	–	8.5.6
XRF	X-ray fluorescence	–	7.2
y	distance between two surfaces	m	2.1.4
z	height of the surface	m	8.3.1
z_i	surface height at position i	m	8.3.1
α	factor for viscosity dependency of pressure	–	2.1.1.4
β	factor for viscosity dependency of temperature	–	2.1.1.3
γ	surface tension	N/m	2.1.3.1
$\partial u/\partial y$	shear rate	1/s	2.1.1.1
η	dynamic viscosity	Pa s	2.1.1.1, 2.4, 5.5.2
η_0	dynamic viscosity at atmospheric pressure	Pa s	2.1.1.4
η_{40}	dynamic viscosity at 40°C	Pa s	2.1.1.3
$\eta_{initial}$	dynamic viscosity before test	Pa s	2.1.1.5
η_{after}	dynamic viscosity after test	Pa s	2.1.1.5
$\eta(p)$	dynamic viscosity at pressure p	Pa s	2.1.1.4
$\eta(T)$	dynamic viscosity at temperature T	Pa s	2.1.1.3
$\Delta\eta_{rel}$	permanent shear loss	Pa s	2.1.1.5
θ	fraction of sites occupied by adsorbed molecules	–	4.1.6
λ	thermal conductivity	W/m, K	2.1.4
Λ	film parameter	–	1.2.2, 5.2.1
μ	coefficient of friction	–	1.2.1, 1.2.2, 6.4
ν	kinematic viscosity	m^2/s	2.1.1.2, 2.4
ρ	density	kg/m^3	2.1.1.2
$\Delta\rho$	difference in density	kg/m^3	2.1.3.1
τ	shear stress	Pa	1.2.1, 2.1.1.1

List of Tables

1.1	Typical friction coefficients for different types of contact conditions	10
1.2	The NLGI consistency number of greases	16
2.1	Standard methods for lubricant properties characterization	43
3.1	Some properties of different hydrocarbon chemical structures	47
3.2	Classification and properties of the base fluids, as presented by API and ATIEL	48
3.3	Origin and description of different base fluids	49
3.4	Typical composition of crude oils	50
3.5	The influence of crude oil processing on some base oil properties	52
3.6	The properties of mineral base fluids	60
3.7	The properties of vegetable oils and esters. Values have been approximated for a range of monoesters, diesters and polyolesters	61
5.1	The viscosity classification according to ISO shown as ISO VG (i.e. viscosity grades; KV_{40}, kinematic viscosity at 40 °C)	103
5.2	Viscosity classification of automotive gears according to SAE J306 (KV_{100}, kinematic viscosity at 100 °C)	106
5.3	The viscosity classification of engine oils according to SAE J300. W (as in winter) is used for low temperatures	109
6.1	Summary of possible types of contact areas and motion in the presented tribotests	132
7.1	Summary of used oil analyses performed on hydraulic oils, gear oils and engine oils	147
8.1	Some methods for microscopy, surface measurement and hardness measurement, ordered by scale of detail in information	150
8.2	Selected surface analysis methods, with type of beam used for irradiation and type of signal detected	165
8.3	Summary of performance parameters of the presented microscopy and surface measurement methods. All values are approximate and may depend on operating conditions and the analysed material	180
8.4	Summary of performance parameters of the presented surface analysis methods. All values are approximate and depend on instrument, operating conditions and the analysed material	181

Part One

Lubricant Properties

1

Introduction to Tribology

When surfaces move relative each other, friction and wear arise. Tribology is the science and technology of interacting surfaces in relative motion and covers the three subjects of friction, wear and lubrication. Practical examples of tribology are found anywhere. This can be a slippery shoe on ice or a piece of soap in a wet hand, but also a well lubricated smoothly running machine or mechanical system.

Baldos, a unique car built by students at Luleå University of Technology, is an excellent example of development of new machines with a focus on design for low energy losses (see Figure 1.1). Baldos has been invited to different worldwide events in, for example, Washington DC, Stuttgart and Monaco.

The term 'tribology' is derived from the Greek word *tribos*, which means rubbing [1]. *Friction, wear and lubrication* can be treated separately, each of them with its own underlying questions, but in order to understand tribology a system approach is required [2–5].

The subject of tribology is visualized in Figure 1.2. Friction and wear can be investigated on their own in unlubricated or lubricated contacts. Lubrication is supplementing the areas of friction and wear. In this book, examples from hydraulics, gears and combustion engines are used.

Tribology is highly important for sustainable growth of industry and society. High levels of friction and wear will lead to high energy consumption and in the worst case will shorten the life of the system. Sustainable growth implies, for example, reducing consumption of raw materials and toxic or environmentally harmful surface materials and lubricants. This means that even minor reductions in friction and prolonged life of lubricated systems yield large monetary and environmental savings.

The impact of environmental demands is exemplified by the emission legislation placed on the automotive segment regarding emissions of both particulate matter and greenhouse gases. Improved tribological performance can make vehicles more energy efficient. Both reduction in friction and improved wear resistance have significant effects on fuel consumption and thus also on greenhouse gas emissions. Good tribological performance is therefore of great importance in order to meet the increasingly tougher demands for emissions from cars and trucks.

Lubricants: Introduction to Properties and Performance, First Edition.
Marika Torbacke, Åsa Kassman Rudolphi and Elisabet Kassfeldt.
© 2014 John Wiley & Sons, Ltd. Published 2014 by John Wiley & Sons, Ltd.

Figure 1.1 Baldos – a machine with low energy losses

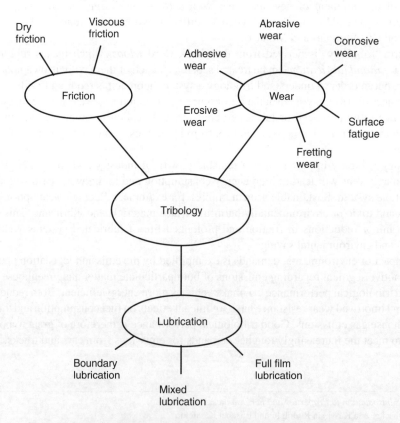

Figure 1.2 The different areas of tribology: friction, wear and lubrication

Introduction to Tribology

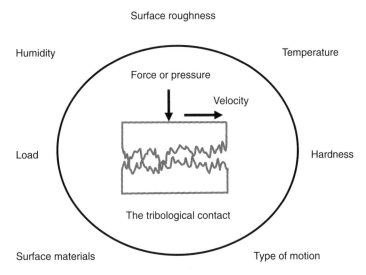

Figure 1.3 Tribological contacts are affected by different conditions

General trends in the development of products and processes aim at higher power density, reduced friction losses, longer wear life, reduced lubricant consumption, less toxic or environmentally damaging lubricants and additives, reduced use of environmentally harmful surface treatment processes and reduced weight of components.

This chapter will introduce the reader to the most important tribological phenomena, covering the tribological contact followed by friction, wear and lubrication.

1.1 Tribological Contacts

A tribological contact is defined as two solid bodies in contact under relative motion. It can be either unlubricated or lubricated. The tribological contact is characterized by its operating conditions (e.g. velocity, load and type of motions), material parameters (e.g. surface material, surface roughness and hardness), environmental conditions (e.g. temperature and humidity) and, in the lubricated case, lubricant properties (e.g. viscosity) (see Figure 1.3).

The tribological contact can be observed at different scales, that is at macroscopic scale (or macroscale) or at microscopic scale (or microscale) (see Figure 1.4). The *macroscale* will give global information of the contact, while the *microscale* will give local information within the

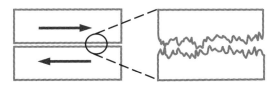

Figure 1.4 The tribological contact can be observed at macroscale (left) or at microscale (right). Surfaces appearing smooth at macroscale still show roughness at microscale

contact. For example, a contact that appears smooth at macroscale may appear very rough and uneven at microscale. The real contact area between the surfaces is the sum of a large number of small areas where surface peaks from the two surfaces get into contact. As a consequence, the apparent contact area at the macroscale is much larger than the real contact area between the two surfaces in contact.

In engineering systems, as machines, the solid bodies are usually made of steel. Consequently, most contacts are steel–steel contacts. However, yellow metal (e.g. bronze), polymeric materials and ceramics are also used. In these cases the contact is often steel–yellow metal, steel–polymeric material or steel–ceramics. Steel–steel contacts almost always require lubrication while the other examples may be run either unlubricated or lubricated.

1.1.1 Macroscale Contacts

The shape of the bodies in contact determines the overall geometry of the contact, which strongly influences the operating conditions and performance of the contact [6]. A contact is said to be *conformal* if the surfaces of the two bodies fit exactly or closely together, such as two flat surfaces in contact. If the two bodies have dissimilar curvatures the contact is said to be *nonconformal*, such as a curved body mating a flat surface. When such bodies are brought into contact they will touch first at a point or along a line (see Figure 1.5). For example, the contact between the rotating shaft and a bushing is conformal, while the contact between two gear teeth is nonconformal.

At macroscale the contact areas can be categorized as distributed, line or point contact areas (see Figure 1.6). In the *distributed contact area* the load is distributed over an area that extends in both the x and y directions. Such contacts are found in journal bearings, cylinder/piston systems and many other sliding applications. *Line contact* occurs when the profiles of the

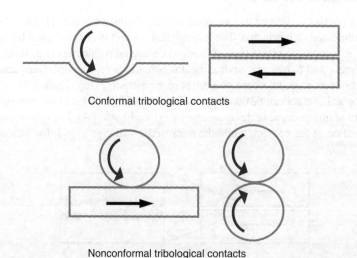

Figure 1.5 Examples of conformal and nonconformal tribological contacts

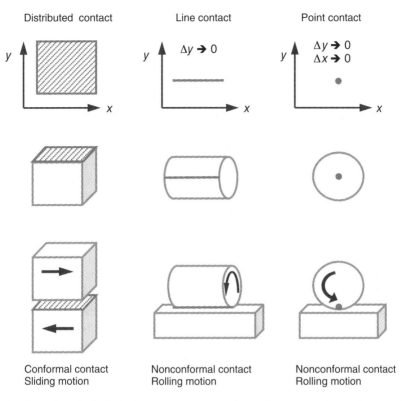

Figure 1.6 Examples of tribological contacts at macroscale are shown according to the type of contact area (For a colour version of this figure, see the colour plate section)

bodies are conformal in one direction and nonconformal in the perpendicular direction, for instance when a cylindrical roller mates a flat surface (e.g. in axial roller bearings). A *point contact* occurs, for example, when a ball mates a flat surface.

Conformal contacts give rise to distributed contact areas, while *nonconformal* contacts give rise to line or point contact areas. However, in reality both point and line contacts always have some extension in both the x and y directions.

The relative motion may be sliding, rolling or a combination of both. *Pure sliding* often occurs in conformal contacts as for a rotating shaft carried by a bushing (or in skiing and skating). The most typical example of *rolling motion* is found in ball bearings. (Rolling may also dominate for cylindrical solid bodies in a rotating motion against a flat surface, as is the case of tyres on a road.)

More complex application geometries may give rise to *combined sliding and rolling* motions. This is the case, forexample, for the relative motion between gear teeth. When both sliding and rolling exist it is common to define the ratio between the magnitudes of sliding and rolling motions.

At macroscale, the contact load appears to be distributed over the whole apparent contact area. It can therefore be characterized by a mean load or it can be transformed into a contact pressure distribution. Often, the contact pressure is high enough to deform the contacting

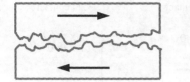 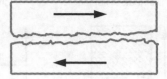

Figure 1.7 The running-in process smoothens rough surfaces

solid bodies. For example, the load in rolling motion with line or point contact may result in a very high contact pressure. When the load is removed, the *elastic* part of the deformation disappears. If a permanent deformation remains, it is referred to as *plastic deformation*.

The load and the relative motion will also give rise to friction in the contact. The friction energy dissipates as heat in the contact, resulting in a temperature rise. At macroscale, the temperature can be characterized by an average value.

1.1.2 Microscale Contacts

At microscale, all technical surfaces in a mechanical contact have a certain roughness. When the two solid surfaces move in relative motion the load will be carried by the asperities (i.e. roughness peaks). This will produce high local pressures and high local temperatures in the individual asperities in contact. Such local conditions often give plastic deformation, which usually makes the surface smoother. At the microscale the chemistry and the mechanical properties of the surface matter. This may include the chemical interaction between the surface material and the lubricant. These local conditions and interactions can be very dramatic, with high local energy dissipation resulting in atomic bond breakage and formation of new atomic bonds.

During the first period of using a new tribological system, wear and plastic deformation of the asperities may occur. This is known as the running-in process, which results in large changes at microscale that are hardly visible at macroscale. Due to surface smoothing, running-in results in reduced local pressure and temperature (see Figure 1.7).

1.2 Friction

Friction is the force resisting the relative motion between two surfaces in contact. It is commonly divided into *dry friction* and *viscous friction*. It may be *static*, that is the solid bodies have no relative motion, or *dynamic*, when the solid bodies are moving relative to each other. Dry friction occurs between two dry solid bodies. Viscous friction occurs when the two solid bodies are more or less separated by a fluid, for example a lubricant.

1.2.1 The Coefficient of Friction

The coefficient of friction μ is defined as the ratio of the friction force F and the normal force N between the bodies, as shown in Figure 1.8 and expressed by

$$\mu = \frac{F}{N} \tag{1.1}$$

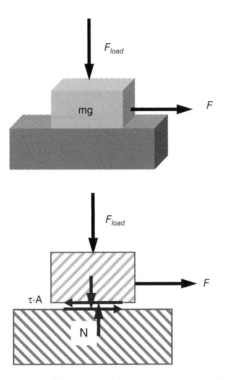

Figure 1.8 Friction visualized as pulling a small box across a flat surface. The set-up (top) and a cross-section view (bottom). The forces at the interface also are shown

where the normal force is actually the sum of the load of the mass mg and any externally applied load, F_{load}:

$$N = mg + F_{load} \qquad (1.2)$$

The normal force always acts perpendicular to the contact area. When sliding the mass the friction force is the force required to maintain the sliding and it always acts tangential (i.e. parallel) to the contact area.

In tribological contacts external load is often applied to magnitudes making the mass weight negligible, that is $mg \ll F_{load}$, which implies

$$\mu = \frac{F}{F_{load}} \qquad (1.3)$$

The coefficient of friction is obtained from known or measured forces. Both F and F_{load} can be measured with use of, for example, a load cell and may thus be logged as a function of time.

For a contact where the surfaces are fully separated by a lubricant the friction force is commonly expressed by

$$F = \tau A \qquad (1.4)$$

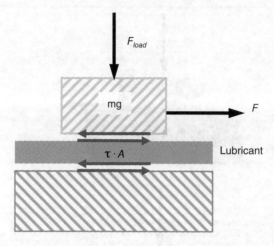

Figure 1.9 Illustration of shear forces between solid bodies and fluid in a lubricated contact

where τ is the shear stress in the lubricant and A is the contact area. The shear stress is determined by the lubricant properties, the velocity of the motion and the distance between the bodies. The forces in the lubricated contact act on both the solid bodies and the fluid, as shown in Figure 1.9.

Some examples of friction coefficients are given in Table 1.1. It is apparent that lubricants significantly lower the friction coefficients in sliding contacts. In rolling contacts lubricants are used to distribute the load. This will increase the coefficient of friction because the lubricant hinders the motion of the rolling body through viscous shear. Still, rolling contacts generally yield lower friction than sliding contacts.

1.2.2 Lubrication Regimes

A general description of the friction behaviour in a lubricated contact can be seen in Figure 1.10, where the dependence of the coefficient of friction μ versus the film parameter Λ is shown. The film parameter Λ is calculated as

$$\Lambda = \frac{h}{\sqrt{R_{qA}^2 + R_{qB}^2}} \quad (1.5)$$

Table 1.1 Typical friction coefficients for different types of contact conditions

Case	Friction coefficient	Type of motion
Unlubricated journal bearing	0.1–0.2	Sliding
Oil lubricated journal bearings	0.01–0.09	Sliding
Oil lubricated roller bearings	0.001–0.005	Rolling
Unlubricated roller bearings[a]	0.0005–0.001	Rolling

[a] Only valid for low loads

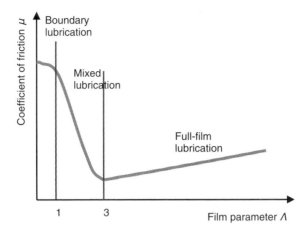

Figure 1.10 Coefficient of friction μ versus film parameter Λ in lubricated sliding contacts

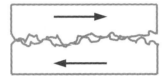

Figure 1.11 Boundary lubrication, where there are always some asperities in contact

where h is the lubricant film thickness and R_{qA} and R_{qB} represent the surface roughness of the two surfaces A and B in contact (see the definition of roughness in Section 8.3). The contact is classified as *boundary*, *mixed* or *full film* lubricated depending on the degree of mechanical contact between the solid surfaces. The curve in Figure 1.10 originates from the Stribeck curve[1] [7].

Boundary lubrication implies heavy contacting between the asperities with a Λ-value below 1. The load is carried by the solid surfaces in contact (see Figure 1.11). The lubricant is mainly acting as a carrier of additives. The presence of additives is necessary to ensure the performance and build-up of a boundary film. This regime is characterized by high load and low speed. A slowly rotating shaft and a bushing mainly work in the boundary lubrication regime even if they are lubricated.

In *mixed film lubrication* the surfaces are less separated than in the full film regime. The surfaces are close enough for asperity contact to occur occasionally. The mixed film lubrication regime is a combination of full film lubrication and boundary lubrication with Λ-values between 1 and 3. Thus, the load is carried partly by a pressure in the fluid film and partly by the asperities in contact, as shown in Figure 1.12. The lubricant will support the contact with necessary additives to reduce wear.

---

[1] In the Stribeck curve the coefficient of friction μ is plotted versus the ratio of velocity and viscosity to load. Since velocity, viscosity and load are easier to measure and control than the film parameter Λ, the Stribeck curve is often used to, for example, present results from laboratory testing.

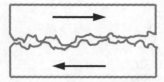

Figure 1.12 Mixed lubrication, where the surface roughness is in the same order of magnitude as the film thickness

A journal bearing working under various speed conditions can alter between the full film lubrication and mixed lubrication.

In *full film lubrication* the solid bodies are lubricated by a thick enough lubricant film to ensure full separation of the surfaces (see Figure 1.13). In this regime the coefficient of friction is very low. A Λ-value higher than 3 indicates full film lubrication. Hydrodynamic bearings are examples of machine components carefully designed to operate in the full film regime.

Elastohydrodynamic (EHD) *lubrication* is part of the full film regime. In this regime the load is high enough to cause elastic deformation of the surfaces. Elastohydrodynamic lubrication can be found in nonconformal contacts, such as a ball on a flat surface when the motion is rolling or rolling and sliding. Performance in the EHD conditions is mainly governed by the viscosity and the viscosity–pressure dependence.

The *lubricant film thickness h* (see Equation (1.5)) is determined by the lubricant properties (see Chapter 2), the operating conditions, the contact geometry and the solid surface's material properties [8]. In practice, typical lubricant film thicknesses are about 1–100 μm.

1.3 Wear

Wear is loss of material from a solid surface. Wear can appear in many ways depending on the material of the interacting contact surfaces, the environment and the operating conditions. At least five principal wear processes can be distinguished: abrasive wear, adhesive wear, surface fatigue and fretting and erosive wear. These are briefly described below.

Adhesive wear occurs under sliding conditions where asperities are plastically deformed and welded together by high local pressure and temperature. When the sliding continues, the asperity bonds are broken and the result is removal of material or transfer of material from one surface to the other. Extensive adhesive wear is commonly described as scoring, galling or seizure.

Abrasive wear may occur when a rough hard surface mates a softer material where the asperities of the hard material scratch the softer surface. This process is called two-body

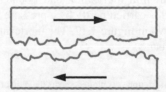

Figure 1.13 Full film lubrication, where the surface roughness is much smaller than the film thickness

abrasive wear. If hard loose wear particles are present between the mating surfaces one or both surfaces can be worn by scratching. This situation is called three-body abrasive wear.

Surface fatigue occurs when cyclic loading weakens the material and can be the predominant wear mechanism in rolling contacts involving some sliding. This may result in subsurface cracks that may propagate and lead to material losses. Surface fatigue is sometimes also called pitting when small pieces of material break away from the surface, forming pits.

Fretting wear, or fretting, occurs when there is a very small oscillatory relative motion between two surfaces in contact. Often the intention is that the surfaces should be fixed, but due to, for example, vibrations some motion still occurs. The term fretting is often used to denote damage mechanisms such as fretting fatigue, fretting wear and fretting corrosion.

Erosive wear occurs in situations where hard particles impact a solid surface and remove material. The impacting particles can be stones, ore pieces, or small particles carried in a water or air jet.

Often more than one wear mechanism is active at the same time. For example, small, abrasive particles may be generated due to adhesive wear or surface fatigue, which may also lead to abrasive wear. In addition, *tribochemical wear* can occur, which involves chemical reactions between the solid surfaces and surrounding lubricant or environment. The chemical reactions, such as corrosion, can weaken the surface layer, which will enhance the effect of other wear mechanisms.

1.3.1 Wear Rate

Wear is often classified as mild or severe from an engineering point of view. *Mild wear* often results in a surface that is smoother than the original surface. On the other hand, *severe wear* often results in a surface that is rougher than the original surface. Wear is an ongoing process between two mating contacts that has to be controlled. Typical wear behaviour is shown in Figure 1.14. Starting with newly manufactured surfaces, the process starts with a running-in

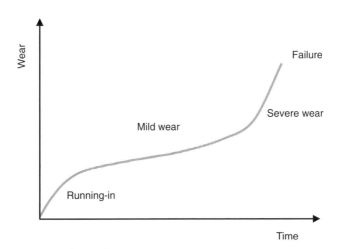

Figure 1.14 The wear process depends on time with periods of running-in, mild wear, severe wear and failure

period, followed by, if correctly designed, a mild wear period. The life of the components ends if the wear rate significantly increases and causes failure.

A severe wear situation is not acceptable in most cases. Mild wear is what the engineers strive for. It can be obtained by proper shape and surface roughness of the contact surfaces together with proper materials. However, often lubrication of the surfaces is necessary to secure mild wear.

Wear can be expressed as wear volume or wear rate. The wear rate is commonly expressed by Archard's wear model

$$\frac{V}{s} = k_w N \qquad (1.6)$$

where V is the volume removed from the surface, s is the sliding distance, k_w is the specific wear number and N is the normal load. The result is a ratio often called the wear rate, given as a number in units of mm^3/m [9].

1.4 Lubrication of the Tribological System

When designing a mechanical system it is obviously important that the designer from the very beginning considers the properties and performance of all parts of the system, including the lubricant. Analogously, the designer of the lubricant must foresee the operating conditions and environment for it in order to formulate the optimum combination of base fluids and additives for each application (see also Chapter 5).

In this section, the purposes of lubricating a tribological system are presented. Lubricant properties and characteristics of base fluids and additives are presented in the following chapters (Chapters 2, 3 and 4).

1.4.1 The Purposes of Lubricants

Lubricants can serve several functions in tribological contacts. The overall purpose of it is to control friction and wear. However, the diversity of the lubricant function comprises the following:

- separate moving parts
- transfer heat
- transfer power
- reduce friction
- protect against wear
- prevent corrosion
- carry away contaminants and debris
- seal for gases
- reduce noise and vibrations.

It is crucial to separate the moving parts in order to reduce friction and avoid wear. The relation between the lubricant film thickness h and the surface roughness R_q shows the ability of the lubricant to separate the surfaces (see Equation (1.5)). In order to separate the surfaces fully, that is to transfer the contact into the full film lubrication regime, the lubricant film thickness has to exceed a minimum value.

The friction that occurs in the contact will raise the local temperature. The temperature rise may be high enough to soften or to almost cause local melting of the material. Thus, a very important lubricant property is to transfer the heat away from the contact. Important thermal properties are the specific heat capacity and the thermal conductivity.

Lubricants are in some applications used to transfer power. This is done, for example, in hydraulic systems where power is transferred from one point to another with the aid of the lubricant.

When two surfaces in a sliding contact move close to each other, separated by a thin lubricant film, friction will result in the contact. The friction would be much higher without the lubricant. This friction-reducing property is commonly referred to as *lubricity*, which is determined by tests since it cannot be measured directly.

Lubricants protect against wear just by being in the contact, since it reduces friction and local temperatures. They are also formulated to reduce wear by adding antiwear and extreme pressure additives.

Lubricants protect against corrosion by covering the surfaces and protecting them from, for example, oxygen and water. In addition, they are formulated to reduce corrosion even further by adding corrosion inhibitors.

It is important to reduce the amount of particles in the contact interface. Hard particles may give rise to abrasive wear in the contact even if a proper lubricant has been selected. Metal particles may also catalyse oxidation and consequently cause ageing. The lubricant will act as a carrier and will transfer contaminants as wear debris and particles from the contact. It is also common to have a filter in the application that collects contaminants and debris. The lubricated contact is usually narrow. At the narrow opening the capillary forces will efficiently seal for leakage of gases and liquids. In many applications the use of lubricants will also reduce noise and vibrations that originate from the contact or are transferred through the contact.

1.4.2 Reducing Friction and Protecting against Wear

A lubricant can reduce wear, either directly or by reducing the friction. A lubricant can also increase wear, for example by chemical attack or by being a trap for abrasive particles causing abrasive wear or fatigue.

Lubrication is particularly effective in reducing adhesive wear of sliding contacts. Also, chemical wear may be reduced by lubrication, if, for example, the lubricant prevents contact with a corrosive environment. On the other hand, fluid lubricants are generally not very effective for reducing other wear mechanisms, such as abrasive wear, erosive wear or fretting wear. Finally, when adhesive wear is reduced by lubrication, asperities may remain, causing enhanced surface or pitting fatigue.

Lubricants have different roles in different lubrication regimes (see Figure 1.10). In the boundary lubrication regime the surfaces are in contact with each other and the main purpose of the lubricant is to carry the additives into the contacts. The additives form a film on the surfaces, which reduces the friction and wear. If the operating conditions change, a situation may be reached where the additive film cannot form properly and severe adhesive wear may occur.

Table 1.2 The NLGI consistency number of greases

NLGI group number	Description	Penetration (10^{-1} mm)
000	Fluid	445–475
00	Semi-fluid	400–430
0	Very soft	355–385
1	Soft	310–340
2	Normal grease	265–295
3	Firm	220–250
4	Very firm	175–205
5	Hard	130–160
6	Very hard	85–115

In the full film lubrication regime the surfaces are completely separated by the lubricant film, indicating that no wear should occur. The main purpose is to separate the surfaces and minimize friction.

Also, in a component designed for operation in the full film regime wear may happen. It may occur during, for example, starting and stopping, before a sufficiently thick lubricant film has been formed. This means that during a short period of time the contact will act in the boundary and mixed lubrication regimes. For a new component, that is new surfaces, the smoothening of the surfaces during the running-in process enhance the possibilities to achieve full film lubrication.

1.4.3 Semi-Solid Lubricants

Greases are used when liquid lubricants cannot be used, for example when a liquid lubricant will not stay in the contact or for sealing purposes to protect the contact from contamination by debris or water or a corrosive environment. Greases are semi-liquids consisting of base fluid with 5–30% thickener. The base fluid type can be varied (i.e. mineral, synthetic or vegetable oil). Using a synthetic base fluid will improve low and high temperature properties of the grease. The thickener can be soap, a polymer or clay. The base fluid lubricates the contact, while the thickener allows the grease to remain in the contact and protect against contamination.

Greases are described by their consistency, dropping point and base fluid viscosity (refer to Chapter 2). Consistency is measured in NLGI[2] consistency numbers (see Table 1.2) [10]. It is measured by allowing a cone to sink into the grease measuring the penetration depth. Normal greases have a consistency number of 2.

The dropping point is the temperature at which the grease changes phase from semi-solid to fluid. Different greases will melt at different temperatures depending on the thickener and the interaction of the base fluid and the thickener.

1.4.4 Solid Lubricants and Dry Lubricants

Lubrication can also be achieved through *solid lubricants* [11]. Solid lubricants are used when there are problems with keeping a fluid lubricant in the contact, component or system, for

---

[2] National Lubricating Grease Institute.

example when the contact pressure or temperature becomes too high for a fluid lubricant to function properly. Different types of solid lubricants are graphite, molybdenum disulfide and PTFE (polytetrafluoroethylene. They may be applied as coatings or as solid additives in powder form carried in greases.

Low friction *coatings* are also used in combination with a fluid lubricant, mainly to protect the surfaces in the boundary lubrication regime. The most commonly used type of low friction coating is DLC diamond-like-carbon coatings.

Soft material coatings, such as lead- or tin-based Babbitt coatings, are also used in order to secure the operation at boundary lubrication. The manufacturing process and costs must be taken into account when choosing the type of coating.

Semi-solid and solid lubricants will not be further explored in this book.

References

[1] Jost, P. (ed.) (1966) *Lubrication (Tribology) Education and Research* ('Jost Report'), Department of Education and Science, HMSO.
[2] Blau, P.J. (ed.) (1992) *Friction, Lubrication, and Wear Technology*, vol. 18, ASM Handbook, ASM International.
[3] Bhushan, B. (ed.) (2001) *Modern Tribology Handbook*, vols 1 and 2, CRC Press.
[4] Jacobson, S. and Hogmark, S. (2011) Tribologi–Friktion, Nötning, Smörjning, available from Uppsala University.
[5] Hutchings, I.M. (1992) *Tribology, Friction and Wear of Engineering Materials*, Metallurgy and Materials Science Series, Edward Arnold.
[6] Johnson, K.L. (1985) *Contact Mechanics*, Cambridge University Press.
[7] Stribeck, R. (1902) Die wesertlichen Eigenschaften der Gleit- und Rollenlager. *Zeitschrift des Vereines Deutscher Ingenieure*, **36**(46), 1341–1348, 1432–1438, 1470.
[8] Hamrock, B. and Dowson, B. (1981) *Ball Bearing Lubrication: The Elastohydrodynamics of Elliptical Contacts*, John Wiley & Sons, Inc., New York.
[9] Archard, J.F. (1953) Contact and rubbing of flat surfaces. *Journal of Applied Physics*, **24**(8), 981–988.
[10] Boehringer, R.H. (1992) Grease, in *Friction, Lubrication, and Wear Technology*, ASM Handbook, vol. 18 (ed. P.J. Blau), ASM International.
[11] Erdemir, A. (2001) Solid lubricants and self-lubricating films, in *Modern Tribology Handbook*, vol. 2 (ed. B. Bhushan), CRC Press.

2

Lubricant Properties

The lubricant properties are important for a satisfactory function of the application. Since the purposes of lubrication vary between applications, the demands on the lubricant also vary. Therefore, there are many different ones available. The lubricant properties are verified with analyses and testing, both during development and by quality control of new and used oils. In this chapter different lubricant properties are described and for each property an example of an analysis method is given. Lubricant characterization will be covered in Chapter 7. When applicable, this chapter covers the following structure for each selected property of liquid lubricants:

- general importance
- importance for the lubricated contact
- definition
- method[1] of measuring and conditions for the measurements.

The lubricant properties can be divided into performance properties, long life properties and environmental properties [1, 2] (see Figure 2.1). These properties are placed in ovals in the diagram, stretching from an initial time to a final time, which differs.

The *performance properties* are central from the very instant the lubricant is added to the contact. These properties make sure that the purposes for lubricating the tribological contact are fulfilled (see Section 1.4). These properties include, for example, viscosity, thermal properties, low and high temperature properties, and air and water contaminant sensitivity.

The *long life properties*, or the longevity, will prolong the functionality of the lubricated contact. These properties involve aspects that enhance the life of the lubricant and the lubricated contact, thus minimizing the risk of failure and need of service. Long life properties involve oxidation stability, hydrolytic stability and corrosion inhibition. In addition, storage and handling will be covered since they provide the basis for lubricant longevity.

---

[1] There are sometimes several methods available. However, for simplicity reasons one or two are chosen as a reference. These are tabulated at the end of this chapter.

Lubricants: Introduction to Properties and Performance, First Edition.
Marika Torbacke, Åsa Kassman Rudolphi and Elisabet Kassfeldt.
© 2014 John Wiley & Sons, Ltd. Published 2014 by John Wiley & Sons, Ltd.

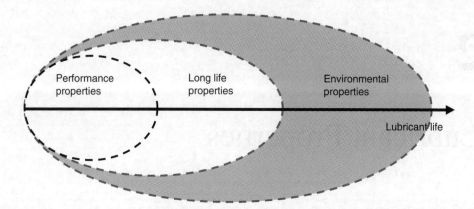

Figure 2.1 The timescale of different lubricant properties (For a colour version of this figure, see the colour plate section)

The *environmental properties* stretch to long times. They may have an acute influence on either people or the environment during, for example, immediate human exposure, oil spills and leakages to sensitive environments. They may also have an influence for a very long time in humans or on the environment if the lubricant is based on, for example, bioaccumulative substances. Environmental properties include both working environment and properties such as biodegradability, renewability, toxicity and bioaccumulation. Environmental demands increase continuously year by year and cover pollution to air, land and sea.

2.1 Performance Properties

Performance properties make sure that the purposes of the lubricant are fulfilled. For instance, separating moving surfaces will be possible by having the right lubricant film thickness governed by the viscosity. The temperature characteristics, such as volatility, flash point and pour point, set the possible operating temperature range. The air and water entrainment properties, such as foaming, air release and emulsibility, determine whether there is a lubricant film in the contact or not. One could argue whether these latter properties should be mentioned as performance properties or long life properties. However, they are mentioned here due to their possible negative effect on the lubricated contact as well as their impact on the lubricant itself.

Measurements and analyses of performance properties should be done under as realistic application conditions as possible. Therefore, most analyses are performed at application temperatures and the laboratory environment should ensure control of important parameters. Performance properties are all physical properties. This implies that they can be measured without changing the chemical identity of the lubricant.

2.1.1 Viscosity

The most important performance property is viscosity. The viscosity determines the lubricant film thickness and thereby the performance of the lubricated contact [1]. In general, a high viscosity will give a thicker lubricant film, while a low viscosity will give a thinner film. If the lubricating film is too thin, asperities will be brought in contact with each other and friction will increase. On the other hand, if the film thickness is too large, more energy is required to move the surfaces. In addition, since pumpability decreases with increasing viscosity, a lubricant with excessively high viscosity may in the worst case not be possible to pump to the contact, resulting in starvation and consequently wear or seizure.

The viscosity changes with temperature, pressure and shear rate. All these aspects are dealt with during the development and lubricants are available in different viscosity grades to satisfy the requirements in different applications. Generally, the lowest viscosity lubricant that still forces the two moving surfaces apart is desired.

2.1.1.1 Dynamic Viscosity

Viscosity is defined as the resistance to flow of a fluid deformed by shear stress. It describes a fluid's internal resistance to flow and may be thought of as a measure of fluid friction. A fluid lubricant between two horizontal plates, or solid bodies, one fixed and one moving with a constant speed, can be used to illustrate viscosity (see Figure 2.2). The fluid can be seen as consisting of layers moving at different velocities, where close to the stationary body the lubricant velocity is equal to zero and close to the moving body the lubricant velocity is equal to the moving body's velocity.

The shear stress τ, required to move the upper moving solid body at a constant velocity u, is defined by Newton's law to be

$$\tau = \eta \frac{\partial u}{\partial y} \tag{2.1}$$

where u is the velocity of the moving solid body and y is the distance between the two solid bodies. The change in velocity with respect to height, the term $\partial u/\partial y$, is known as the shear

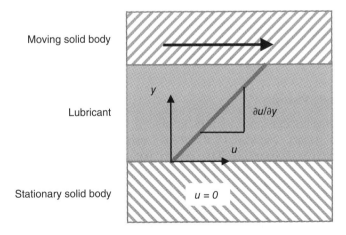

Figure 2.2 A linear velocity profile of the lubricant layer. Close to the stationary body the lubricant velocity is equal to zero and close to the moving body the lubricant velocity is equal to the moving body's velocity

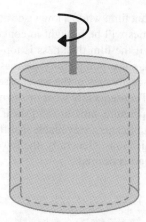

Figure 2.3 Dynamic viscosity is measured using a rheometer. Fluid is filled in the space between an inner rotating cylinder and an outer fixed cylinder and the torque required to keep the outer cylinder fixed is measured

rate. The dynamic viscosity, η, is the proportionality factor giving the ratio of shear stress to shear rate (see also Equation (1.4) and Figure 1.9).

Many fluids follow the linear Newton's criterion, that is Newtonian fluids where the viscosity is independent of the shear rate. Among these are water and oils, that is a linear velocity profile arises in a lubricated contact in relative motion (see Figure 2.2). If the relationship between shear stress and the velocity gradient (shear rate) is not constant, for example at low temperatures, the behaviour is known as non-Newtonian.

The dynamic viscosity is measured in a rheometer. The liquid is placed in the space between two concentric cylinders (see Figure 2.3). This rheometer is also known as a rotational viscometer. The inner cylinder has a known velocity, the clearance is given and the torque needed to keep the outer cylinder fixed is measured. This gives the shear stress for a given shear rate. Measurements are performed at the temperature of relevance to the application.

The unit of dynamic viscosity is Pa s or the more commonly used cP (centipoise), where 1 mPa s = 1 cP. Water at room temperature has a dynamic viscosity of about 1 mPa s and engine oil has about a couple of hundred mPa s.

2.1.1.2 Kinematic Viscosity

The contact experiences dynamic viscosity. In spite of that, the kinematic viscosity is often used to specify lubricants. One reason is that the kinematic viscosity is easier to measure. The kinematic viscosity v is independent of the density and is defined as

$$v = \frac{\eta}{\rho} \tag{2.2}$$

where η is the dynamic viscosity and ρ is the density.

Kinematic viscosity is measured by allowing the lubricant to flow through a capillary with a defined diameter only affected by the gravity. The time for a certain amount of lubricant to flow thorough the capillary is measured. The set-up is known as a capillary viscometer

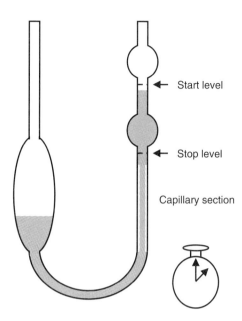

Figure 2.4 Kinematic viscosity is measured in a capillary viscometer. The fluid flows due to gravitation and the time for a certain amount of fluid to flow through the capillary section is measured (For a colour version of this figure, see the colour plate section)

(see Figure 2.4). The kinematic viscosity is often measured at 40 °C (KV_{40}) and 100 °C (KV_{100}), which cover common application temperatures.

The unit of kinematic viscosity is mm^2/s or the more commonly used cSt (centistokes), where $1\ mm^2/s = 1$ cSt. Water at room temperature has a kinematic viscosity of about $1\ mm^2/s$.

For all lubricants the density decreases with increasing temperature. Consequently, the dynamic viscosity and the density should be measured at the same temperature in order to get an accurate measurement of the kinematic viscosity.

Lubricant densities commonly vary between 800 and 900 kg/m^3. The density has little influence in the actual lubricating contact. However, it is important to know the density, for example, in the case of an oil spill to water determining if the lubricant will sink or float [2].

2.1.1.3 Viscosity Dependence on Temperature

Generally, the viscosity decreases with higher temperature, that is the lubricant gets thinner. The viscosity dependence on temperature can be reasonably well described as

$$\eta(T) = \eta_{40} e^{-\beta(T-40)} \qquad (2.3)$$

where η_{40} is the dynamic viscosity at 40 °C and $\eta(T)$ is the dynamic viscosity at the selected temperature T (°C). The β-value is the factor showing the viscosity dependency of the temperature [3]. Typical values are in the range 0.05–0.07 1/°C, depending on the type of base fluid [4]. The exponential relationship implies a strong relation between viscosity and temperature (see Figure 2.5).

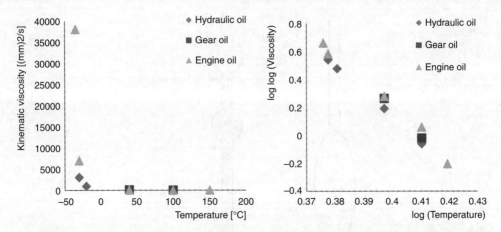

Figure 2.5 The kinematic viscosity dependence on the temperature is shown for hydraulic, gear and engine oils (left) and log log kinematic viscosity as a function of log temperature for the same fluids (right)

This temperature dependency can also be described by the viscosity index (VI, where a higher value means a smaller effect of temperature on the viscosity. The VI is constant for each lubricant, but differs between different lubricants. In Figure 2.6 log log (viscosity) versus log (temperature) is presented for different lubricants. In this type of graph the slope of a curve is related to the VI.

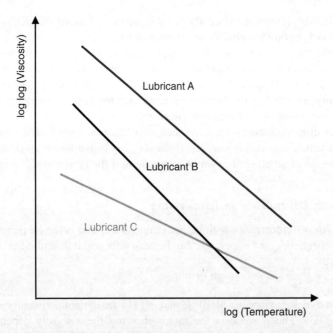

Figure 2.6 Viscosity dependency of temperature for three lubricants is shown in a log log viscosity–log temperature graph. The slope of the curves represents the viscosity index (VI). Lubricant C has the highest VI and lubricant B the lowest

The VI was originally described by choosing the two families of oils with the highest and lowest VI-values at that time. The first family (paraffinic base oils) having a high viscosity index, VI = 100, and the second family (naphthenic base oils) having a low viscosity index, VI = 0. The VI of today's oils may be significantly higher than 100. This is due to use of synthetic base oils, which have a higher inherent viscosity index, and a viscosity index improving additives (refer to Chapters 3 and 4).

A high viscosity index is required in many applications for a lubricant to perform across a wide temperature range. At high service temperatures a sufficiently high viscosity is required to develop a lubricant film. On the other hand, when exposed to low temperatures, the lubricant should not become too thick to allow it to flow.

2.1.1.4 Viscosity Dependence on Pressure

The lubricant viscosity increases with increasing pressure.[2] In most cases the lubricant compressibility is not important, but in some dynamic sequences it may be crucial. The viscosity–pressure dependence is described in the Barus expression as

$$\eta(p) = \eta_0 e^{\alpha p} \tag{2.4}$$

where $\eta(p)$ is the dynamic viscosity at the selected pressure p, η_0 is the viscosity at atmospheric pressure and α is the factor showing the viscosity's dependence on the pressure [5]. At high enough pressure the lubricant will solidify. This may be the case in highly loaded contacts such as ball bearings. Typical values for α are about 24 GPa^{-1} at 40 °C and atmospheric pressure [4].

2.1.1.5 Viscosity Dependence on Shear Rate

The viscosity is independent of shear rate for Newtonian fluids while it decreases with shear rate for non-Newtonian fluids. Lubricants show a Newtonian behaviour at shear rates up to 10^5–10^6 s^{-1}. Low viscosity oils are in general Newtonian due to their low molecular weight. High viscosity lubricants (e.g. emulsions and polymer-thickened lubricants) may show non-Newtonian behaviour due to their complex structure. The most common non-Newtonian behaviour for lubricants is pseudoplastic behaviour, where the viscosity decreases with increasing shear rate, as shown in Figure 2.7.

Pseudoplastic behaviour is due to the alignment of long molecules at increasing shear rates and occurs regularly for multigrade oils containing viscosity index improvers (refer to Chapter 4). The process is in general reversible, which means that when the shear rate is reduced the original viscosity will be retained. With permanent shear loss the molecular structure is broken and the lubricant viscosity will therefore not return to its original value, rendering a permanent shear loss.

There are applications with high shear rates for a long time. Then the viscosity may decrease as a function of time, that is thixotropic behaviour. The phenomenon is referred to as shear

---

[2] The dependence on pressure is sometimes referred to as piezo viscosity in literature.

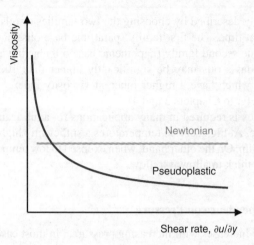

Figure 2.7 The viscosity is constant with the shear rate for Newtonian fluids while it decreases with the shear rate for pseudoplastic fluids

duration thinning. The reduction in viscosity is higher when the lubricant is exposed to higher shear rates (see Figure 2.8).

The shear stability is tested in tapered roller bearings or via diesel injectors (i.e. the Kurt Orbahn test) where the viscosity is measured before, $\eta_{initial}$, and after testing, η_{after}. The permanent shear loss, $\Delta\eta_{rel}$, indicates the shear stability of the lubricant as expressed as

$$\Delta\eta_{rel} = \frac{\eta_{initial} - \eta_{after}}{\eta_{initial}} \qquad (2.5)$$

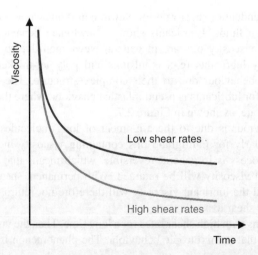

Figure 2.8 Visualization of shear duration thinning. The apparent viscosity is reduced more quickly at high shear rates than at low shear rates

2.1.2 Low and High Temperature Properties of Lubricants

Both low and high temperatures also affect other properties than the viscosity of the lubricant. It may solidify at low temperatures and may evaporate and catch fire at high temperatures. Also, the lubricant may age at increased temperature. However, oxidation stability is covered under long life properties of the lubricant (Section 2.2) [6].

2.1.2.1 Pour Point

The lubricant's pour point is the lowest temperature at which it will flow. The demands differ in the world, for example in the northern hemisphere the requirements can be strong due to the cold temperatures in wintertime. The pour point value is a rough indication of the lowest temperature at which the oil is pumpable. Thus, the lubricant may not reach the contact when approaching the pour point, causing risk of wear or seizure.

The base fluid molecules can move freely in liquid lubricants, but this movement decreases when the lubricant is cooled down. Consequently, both viscosity and density increase when the temperature is lowered. The base fluid molecules will always position themselves in a pattern that is energetically favourable. At a low enough temperature a crystal lattice is formed and the lubricant solidifies.

The pour point can be determined by pouring the lubricant into a beaker, which is gradually cooled down. The beaker is tilted every 3 °C. The pour point is reached when the lubricant no longer flows when the beaker is tilted.

2.1.2.2 Volatility

The lubricant will evaporate when it is heated. Volatility is a measure of the lubricant's tendency to vaporise and depends on its vapour pressure. In closed systems, the vapour is in equilibrium with the liquid. A lubricant with a high vapour pressure will be more readily vaporised than one with a low vapour pressure, at any given temperature. The volatility differs between base fluids. Low molecular weight molecules have a high vapour pressure and will evaporate more easily than high weight molecules. In general, this implies that low viscosity base fluids are more volatile than high viscosity base fluids. Thus, for high temperature applications a balanced mix of base fluids is required for a well-functioning lubricant. Hence, high quality lubricants are formulated with low volatile base fluids.

The amount of lubricant is reduced in the system when it vaporizes. It is important to secure that there is enough lubricant in the contact at any given application temperature. Vaporization also changes the properties of the lubricant when low volatile base fluids are removed. Additionally, the volatility is very important for lubricants used in high temperature applications, where there may be a risk of the lubricant catching fire.

The Noack volatility of a lubricant is measured by heating to 250 °C for 60 minutes in a fume hood (see Figure 2.9). Volatility is evaluated by weighing the lubricant before and after heating. This test is designed for high temperature applications such as engine oils.

2.1.2.3 Flash Point

As was described in the previous section, the lubricant evaporates at high temperatures. At a certain temperature the flash point will be reached. The flash point of a flammable liquid is

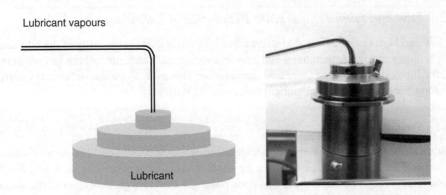

Figure 2.9 Lubricant volatility can be measured using the Noack method. The lubricant is heated and some of the lubricant may evaporate. The weight change is measured

the lowest temperature at which it can form an ignitable mixture in air. At this temperature the vapour may cease to burn when the source of ignition is removed.

The flash point is used to secure good safety and handling. It is important to select a lubricant with a flash point well above the highest application temperature in order to avoid fire. The flash point is primarily important for the total application (i.e. the bulk oil temperature) rather than the local contact.

The most common way of determining the flash point is by heating the lubricant in an open cup (i.e. the Cleveland open cup or COC method). The lubricant is poured into a pan-shaped metallic container and gradually heated (see Figure 2.10). At a certain temperature, that is the flash point, there are enough fumes to ignite. The volatility and the flash point are closely related to each other (see Figure 2.11). In general, lubricants with a high volatility have low flash points.

Figure 2.10 The lubricant is heated during flash point measurements using the Cleveland open cup (COC) method. A source of ignition is brought into contact with the fumes

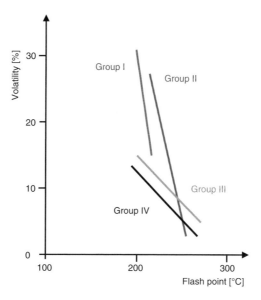

Figure 2.11 The volatility as a function of the flash point for several base oils (see Section 3.2 for definition of base oil groups)

2.1.3 Air and Water Entrainment Properties

Lubricants almost always contain some air and water in different forms. Normally, the lubricants have additives that will secure good air release, low foaming and good demulsibility and thereby good performance in the application (more details in Chapter 4). However, the lubricant may be contaminated by impurities, which may be detrimental to air release (or deaeration), defoaming and demulsibility [7].

2.1.3.1 Air Release

Air occurs in lubricants in the form of free air, foam, entrained air and dissolved air. Free air at the lubricant–air surface may be drawn into the lubricant via, for instance, a pump. This may cause foam and entrainment of air, which increases the oxidation of the lubricant. Dissolved air is invisible and at atmospheric pressure there may be as much as 8% dissolved air in the lubricant. The dissolved air is in equilibrium with the surrounding air at atmospheric pressure, but may be released if the pressure is reduced. This may give rise to disturbances in the system and, in severe cases, cavitation. Poor air release values in hydraulic oils may cause cavitation in the hydraulic pumps and increase compressibility. It is therefore important to have short air release values in such applications.

Air release occurs through air bubble transportation within the lubricant to the air–lubricant surface. The air bubble is formed due to the phenomenon of surface tension γ. The relation between the pressure difference between the air bubble and the surrounding lubricant ΔP, and the bubble radius r, is described by the Young–Laplace equation [8]

$$\Delta P = \frac{2\gamma}{r} \qquad (2.6)$$

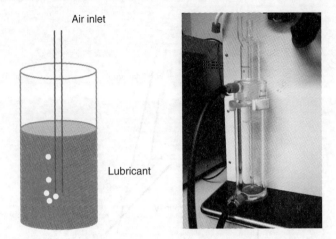

Figure 2.12 Air is bubbled through the lubricant. After bubbling, the time for air release is measured

Air release is enhanced by the growth of air bubbles, due to coalescence or due to a reduction in pressure surrounding the bubble during its rise to the surface. The Young–Laplace equation is applicable to bubbles or droplets formed in lubricants, which is the case in air release, foaming and demulsification.

The rate of bubble rise u_{bubble} can be described by the Stokes equation

$$u_{bubble} = \frac{2g(\Delta\rho)r^2}{9\eta} \qquad (2.7)$$

where g is the gravitational constant, $\Delta\rho$ is the density difference between the air and the lubricant, r is the bubble radius and η is the dynamic viscosity [9]. The radius of the bubble has an exponential effect on the air release rate, that is large bubbles rise quickly while small bubbles may even be entrained in the lubricant. Air release is promoted by increasing the lubricant temperature since this lowers the viscosity.

Air release is measured by bubbling air through a lubricant for 7 minutes at 25 °C, 50 °C or 75 °C (see Figure 2.12). The goal is to test all lubricants at approximately the same viscosity. Thus, low viscosity lubricants are tested at 25 °C and high viscosity lubricants are tested at 75 °C.

After bubbling (aeration) the time for air release is measured. Measuring the density immediately after aeration and repeatedly during the following air release does this. A theoretical value of the density at zero air entrainment is calculated for the lubricant containing 0.2% air. The air release is completed when this level is reached.

2.1.3.2 Foaming

Foaming may occur when air is introduced in a lubricant. Intense foaming may flood the application and consequently reduce the amount of available lubricant.

Foam is defined as a large volume of gas in a small volume of liquid (or solid) material. Pure liquids never foam. There are two criteria to fulfil in order to form foam. First, one component

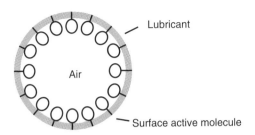

Figure 2.13 The direction of the surface active components in lubricant foam

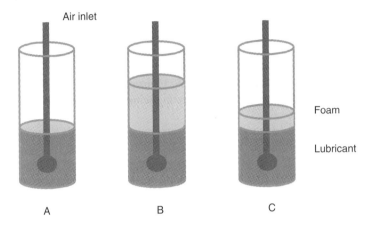

Figure 2.14 Foaming is tested by bubbling air in a lubricant. A: Before bubbling air through the lubricant. B: Test with large foam build-up. C: Test with small foam build-up or significant foam collapse

in the lubricant must be surface active. Second, the surface between air and the lubricant must be elastic. The surface active molecule[3] will arrange itself so that the polar part points towards the air bubble when being in oil (see Figure 2.13).

The foaming characteristics are evaluated at relevant application temperatures, that is room temperature and at an elevated temperature of 94 °C. The foaming is performed by bubbling air at a constant rate through a diffuser, which distributes the air in the lubricant (see Figure 2.14). The foam build-up may be rapid and large or small depending on the lubricant formulation. After 5 minutes the bubbling of air is stopped and the time for the foam to collapse is measured. After 10 minutes the amount of remaining foam, if there is any, is documented. Some foams are extremely resistant to breakage while others break easily and/or rapidly.

Studies have shown that foaming is viscosity dependent for viscosities in the range of 100 to 1000 cSt. The maximum foam volume is observed for lubricant viscosities of 200 to 300 cSt [10]. It is important to cover all relevant application temperatures when evaluating the foaming tendency since the foaming behaviour sometimes differs between temperatures.

---

[3] A surface active molecule has one polar moiety, which is water soluble, and one hydrocarbon chain, which is oil soluble. This will be further discussed in Chapter 4.

In cold climates, it may be well worth preheating the lubricant before usage in order to avoid foaming in cold start-ups.

2.1.3.3 Demulsibility

Water entrainment may result in large or small volumes of water in the lubricant. Large volumes of water usually occur as droplets or a visible water phase, while small amounts of water may exist as dissolved water in the lubricant. In general, droplets are more easily separated from the lubricant than dissolved water. Demulsibility is the ability to separate water (i.e. droplets) from the lubricant. The rate of water separation depends on the combination of base fluids and additives.

It is in general important to have good demulsibility, because water entrainment could cause corrosion, filter problems, wear, and so on. The acceptance of water content differs between applications. Many lubricants used in the industry, such as gear oils, hydraulic oils or turbine oils, are formulated not to allow water to be emulsified in the lubricant. One reason for not accepting water in lubricants is that the operating temperature is below 100 °C. Water is, in general, not a problem in systems with an operating temperature above 100 °C since it will evaporate during operation.

In some applications it is beneficial to allow water entrainment. This is the case for engine oils where water is formed during combustion. If water is separated it may accumulate in the crank case, especially if the car is driven only short distances, not allowing an increase in lubricant temperature. Therefore, dispersant additives (refer to more details in Chapter 4), which disperse small water droplets to a stable emulsion, are added. For these types of lubricants a demulsibility test will not show any water separation at all.

Emulsions may form when water entrains the lubricant. There are two main types of emulsions: oil-in-water (o/w) and water-in-oil (w/o). Water and oil do not naturally mix and emulsions are therefore usually unstable. They can become stable when adding a stabilizer supporting water entrainment (refer to Chapter 4).

Oil-in-water emulsions are most frequent. In emulsion breakage the oil droplets will start to collide and coalesce. The oil droplets will rise in the continuous water phase (also known as creaming). The coalescence proceeds until a continuous oil phase is formed on top of the water phase. The time for the coalescence t_{rc} is determined by

$$t_{rc} = \frac{\Delta \rho r^2}{\eta} \quad (2.8)$$

where $\Delta \rho$ is the density difference between the lubricant and water, r is the droplet radius and η is the dynamic viscosity [9].

Demulsibility is evaluated by blending 40 ml of lubricant with 40 ml of water intensely for 5 minutes at 54 °C. Complete time for the demulsibility is determined when the amount of the remaining emulsion is 3 ml or less (see Figure 2.15).

2.1.4 Thermal Properties

The lubricant inherent thermal properties may be of importance when designing, for example, oil coolers in a lubricated application or calculating temperature in a lubricated contact.

Figure 2.15 Demulsibility is tested by mixing water with the lubricant. A: Test tube before mixing the water and the lubricant. B: Test tube when the water and the lubricant have been mixed. C: Test tube after separation

Important properties are the specific heat capacity and the thermal conductivity. Inherent thermal properties of the lubricants depend mainly on the base fluids and very little on the additives [1].

The *specific heat capacity*, C_p given in units of J/kg K, is a measure of the heat energy required to increase the temperature of a unit quantity of a lubricant. More energy is required to increase the temperature of a lubricant with high specific heat capacity than one with low specific heat capacity. The energy Q required to change the temperature ΔT of a mass m is expressed as

$$Q = mC_p\Delta T \qquad (2.9)$$

For a lubricant with a low C_p value a relatively small energy input will increase the lubricant temperature significantly. Consequently, since the viscosity decreases with higher temperature, the lubricant film thickness will decrease more for lubricants with low specific heat capacity.

The *thermal conductivity*, λ given in units of W/m K, indicates the lubricant's ability to conduct heat, for example how fast heat can be transported away from the lubricated contact. The heat transfer G between two surfaces separated by a lubricant can be described by

$$G = \lambda A \frac{\Delta T}{y} \qquad (2.10)$$

where ΔT is the temperature difference, A is the contact area and y is the distance between the solid surfaces.

In a lubricant between two bodies of different temperatures a temperature profile develops (see Figure 2.16). A large λ-value promotes heat transfer. A large contact area will promote a higher heat transfer from the local contact, thus reducing the temperature difference more quickly.

2.2 Long Life Properties

Long life properties of a lubricant comprise oxidation and hydrolytic stability, as well as corrosion inhibition. The identified long life properties can be referred to as chemical properties,

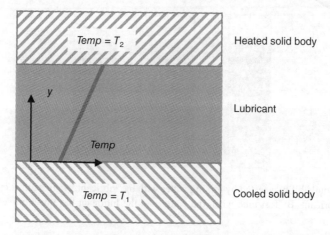

Figure 2.16 A simplified temperature profile in a lubricant between two bodies with different temperatures

since they all involve chemical changes of the lubricant. These properties are investigated by changing the lubricant chemically. Thus, the internal chemical structure of the lubricant must be affected for its chemical properties to be investigated.

It is important to choose the right lubricant for the application in order to prolong the life of both the lubricant and the application. In addition to choosing the right lubricant, it is important to handle and store it in the best way. It is well known that the lubricant ages when exposed to heat, strong light, water, oxygen, metals and impurities. Therefore, it is important to avoid or reduce the exposure during handling and storage.

2.2.1 Total Acid Number (TAN)

The total acid number (TAN) is a measure of the lubricant's condition (or the condition of the base fluid). New (nonused) lubricants may contain acidic components such as organic acids, inorganic acids, esters or some acidic additives. Acids also form when lubricants age. The condition of the lubricant may therefore be evaluated by measuring the acid value. A high TAN usually indicates an aged lubricant. The acid number is one indicator of the remaining life of a lubricant. Monitoring the acid value enables the lubricant to be changed before severe corrosion has occurred in the system.

The acid content, or the total acid number, may be determined by titration. The lubricant sample is diluted in a beaker. Potassium hydroxide (KOH) is added to the sample while mixing. The OH^- from KOH reacts with H^+ from the acidic components in the lubricant, forming water. When all acidic components have reacted with the KOH, the amount of acid is given in mg KOH/g of lubricant, corresponding to how much KOH has been added to reach the equilibrium point. The equilibrium point can be identified by, for example, a colour change.

2.2.2 Total Base Number (TBN)

The total base number (TBN) is another measure of the lubricant's condition and a high TBN indicates a good condition. Nonused lubricants may contain basic additives such as detergents and dispersants, giving a high TBN. These basic additives are added to the lubricant to neutralize acidic components that may form in the lubricant. Therefore, the TBN will gradually be lowered when these additives are consumed, usually followed by oxidation and an increase in viscosity. Consequently, the TBN is a good indicator of the remaining life of engine oils, for instance.

The TBN test is performed by diluting the lubricant sample in a beaker. There are two methods for measuring the TBN. The basic components can either be titrated with a strong acid such as hydrochloric acid or with a weak acid such as perchloric acid. In both cases the acid will react with the basic components. When titrating with perchloric acid both detergents and dispersants can be determined. This is preferred for determining the TBN in new (nonused) lubricants. When titrating with hydrochloric acid, detergents only are determined. This is the preferred method for determining the TBN in used lubricants. When all basic components have reacted with the acid the amount of acid is converted to a KOH equivalent/g of lubricant in analogy with the TAN. The equilibrium point is commonly identified using a potentiometer.

2.2.3 Oxidation Stability

Fresh lubricants have inherent oxidation stability depending on their composition. The lubricant ages by oxidation when it is brought into contact with oxygen at elevated temperatures. The ageing proceeds more quickly at high temperatures according to the Arrhenius relationship

$$k = const \cdot e^{-E_A/RT} \qquad (2.11)$$

where k is the chemical reaction rate constant at temperature T [K], *const* is a constant, E_A is the activation energy required for a chemical reaction to occur and R is the gas constant. The rate of oxidation is doubled every 10 °C and starts to become significant above 60 °C [11].

Oxidation is enhanced by contact with oxygen and metals at elevated temperatures. Thus, oxidation can be lowered by reducing the contact with oxygen (or air), by reducing the temperature in the lubricated contact (or in the application) and by removal of wear debris. In practice, it is difficult to totally avoid the contact with oxygen. However, avoiding mixing air into the lubricant and improved air release will reduce the oxygen content. Cooling the oil will reduce the temperature and filtering will remove wear debris.

Oxidation gives increased acidity and sludge formation, that is the formation of insoluble components. Oxygen consumption, viscosity increase and acid number increase are indications of oxidation occurrence.

Oxidation tests are performed at elevated temperatures in contact with oxygen and metal catalysing the oxidation. Several methods include the addition of water to increase the rate of corrosion during the test. The end of life of the lubricant is determined either by measuring

- the consumption of oxygen
- the total acid number (TAN)

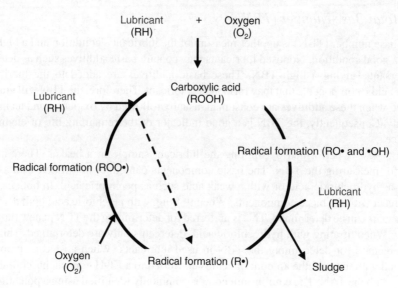

Figure 2.17 The oxidation of a lubricant starts with oxygen contact. It proceeds as a chain reaction process

- the viscosity and
- the consumption of antioxidant additives.

The consumption of oxygen may be measured by comparing an initial pressure of oxygen to a drop in oxygen pressure after a certain time. The TAN, the viscosity and the amount of antioxidants are measured after the oxidation test has been run and the results are compared with initial values for the fresh lubricant. A TAN of 2 mg KOH/g proves that oxidation has proceeded far and is sometimes used to indicate the end of life.

The oxidation process can be explained in more detail (see Figure 2.17). The hydrocarbon chains[4] (RH) in the lubricant react with oxygen during the oxidation process, forming carboxylic acids (ROOH). Oxidation usually starts at the double bonds in the hydrocarbon.

At a certain point the carboxylic acids are cleaved forming radicals, RO• and •OH. A radical is a highly reactive chemical species and is denoted by a point next to the chemical formula. The radicals will immediately react with new hydrocarbons (RH), forming sludge and a new radical, R•. This radical R• will, together with oxygen, O_2, react and form ROO•, which will react with a new hydrocarbon (RH) to form either an R• or ROOH [6].

The oxidation process is a chain reaction, which proceeds until it is stopped by the action of, for instance, antioxidants (refer to Chapter 4).

---

[4] R is used to describe a hydrocarbon chain with an unspecified number of carbons. RH is used to describe hydrocarbons when the hydrogen is involved in a reaction, which is the case in oxidation. This is further described in Chapter 3.

2.2.4 Hydrolytic Stability

Environmentally adapted lubricants are comprised of natural or synthetic esters. These esters may decompose into acids and alcohols when brought into contact with water. This reaction is referred to as hydrolysis where the ester reacts with water, forming acids and alcohol:

$$R_1COOH_2 + H_2O \rightarrow R_1COOH + R_2OH \quad (2.12)$$

The hydrolytic reaction rate is increased at elevated temperatures and is catalysed by metals.

Hydrolytic stability can be evaluated by mixing the lubricant with 10% water at 90 °C for 5 days. It is determined by measuring the total acid number (TAN) where a high TAN increase indicates low hydrolytic stability. A lubricant with low hydrolytic stability should not be used in applications where there is a high risk of water entrainment. Such a surrounding environment may increase the risk of lubricant ageing.

2.2.5 Corrosion Inhibition Properties

Corrosion causes material removal from the metal surface. Severe corrosion may affect the load-bearing capacity of the contact, resulting in machine failure. Lubricant oxidation is also catalysed by corrosion since metal is exposed. Thus, corrosion not only shortens the life of the contact due to surface degradation but potentially shortens the life of the lubricant too.

Corrosion tests are discussed according to the metal being investigated. Thus, the corrosion investigation is divided into steel corrosion (or rust), yellow metal corrosion (or copper corrosion) and soft metal corrosion (commonly referred to as lead corrosion) [6]. Corrosion tests are part of the compatibility studies that are always performed during development of a lubricant formulation.

2.2.5.1 Steel Corrosion

When steel is exposed to water and oxygen (or air) it starts to corrode (commonly referred to as rust). When lubricants contain water, which is more surface active than the lubricant, and oxygen, the corrosion process may start at the steel surface (see Figure 2.18). Iron at the steel surface donates electrons to oxygen, which together with water forms hydroxide ions,

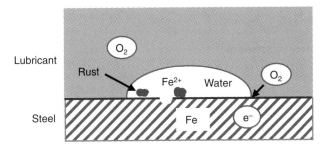

Figure 2.18 In the presence of water and oxygen an iron corrosion process starts, resulting in pitting and rust products

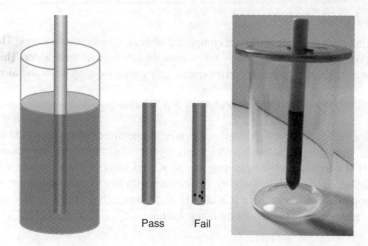

Figure 2.19 Steel corrosion is tested by placing a polished steel rod in the lubricant. The steel rod is evaluated visually after testing

while iron forms iron ions. This happens according to the following reactions (i.e. anodic and cathodic reactions):

$$O_2 + 2\,H_2O + 4e^- \rightarrow 4\,OH^- \quad (2.13)$$

$$Fe \rightarrow Fe^{2+} + 2e^- \quad (2.14)$$

When iron ions are released from the steel surface pitting occurs. The iron ions react with the hydroxide forming iron hydroxide, which precipitates. The hydroxide quickly oxidizes to form rust.

Steel corrosion is evaluated by immersing polished steel rods in the lubricant at an elevated temperature of 64 °C with 10% mix of water (see Figure 2.19). The water can be either distilled or synthetic seawater, depending on application requirements. The corrosion is evaluated and graded by observing the steel rods visually after 4 hours of testing. The grading is no corrosion when the steel rod is not stained, minor corrosion when there are few and small stains and severe corrosion when there are many large rust stains on the steel rod. The test is passed when there is no observed corrosion.

2.2.5.2 Yellow Metal Corrosion

Yellow metals are common in, for example, bearings. Yellow metal may corrode at high temperatures, particularly when brought in contact with certain substances. In a lubricant, substances causing corrosion may be sulfur compounds or oxygen. The corrosion of copper is here exemplified with the reaction with sulfur (see Figure 2.20) and the following reactions:

$$Cu \rightarrow Cu^{2+} + 2\,e^- \quad (2.15)$$

$$S + 2\,e^- \rightarrow S^{2-} \quad (2.16)$$

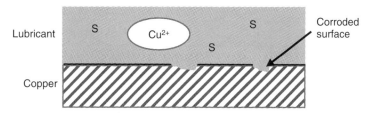

Figure 2.20 Yellow metal corrodes in the presence of sulfur compounds at high temperatures

Electrons are released when copper corrodes. This requires that another substance is available to attract the electrons. In this case sulfur in the lubricant is prone to attract these electrons, forming sulfur ions. The copper and the sulfur ions react, forming copper sulfide. The corrosion may be even across the surface or local giving rise to pits.

Yellow metal corrosion is evaluated by immersing a polished copper strip in a lubricant at an elevated temperature of 120 °C. The degree of corrosion is evaluated by visual inspection after 3 hours (see Figure 2.21). The corrosion is graded according to a given colour scale from 1A, 1B (i.e. minor corrosion), up to 4C (i.e. severe corrosion). The copper strips with minor corrosion are still copper coloured while the copper strips with severe corrosion have turned black.

2.2.5.3 Soft Metal Corrosion

Soft metal corrosion is not part of the standard procedure when evaluating a lubricant. However, soft metals do occur in different bearing materials. Therefore, there is interest in looking into the corrosion of soft metals. In contact with air (or oxygen) the soft metal will oxidize, here exemplified by the oxidation of lead:

$$2Pb + O_2 \rightarrow 2PbO \qquad (2.17)$$

If the lubricant contains esters and water is brought into the lubricant, the ester may hydrolyse, forming fatty acids RCOOH (refer to Equation (2.12)). The soft metal oxide

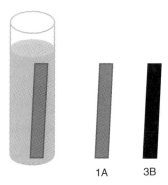

Figure 2.21 Yellow metal corrosion is evaluated by placing a copper strip in a lubricant. Evaluation is graded visually according to a given scale

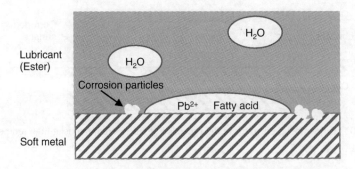

Figure 2.22 Soft metals corrode in the presence of fatty acids and oxygen

(exemplified by lead oxide) may react with the fatty acid, forming a white precipitate and a corroded surface (see Figure 2.22):

$$PbO + 2RCOOH \rightarrow H_2O + Pb(RCOO)_2 \qquad (2.18)$$

Soft metal corrosion can be evaluated by placing a piece of soft metal in the lubricant at elevated temperatures. The weight loss after 168 hours gives the degree of corrosion. The duration of the test can be shorter if severe corrosion is anticipated and longer if corrosion is assumed to be minor. Elemental analyses of the lubricant will supplement the weight measurements of the metal piece in determining the corrosion rate (refer to Chapter 7).

2.3 Environmental Properties

Modern lubricants are relatively harmless to most users, but as with all chemical products they should be handled with care. It is important to wear protective clothing when handling lubricants to avoid skin contact. This is particularly important when handling low viscosity lubricants, both due to risk of uptake via the skin and also due to risk of inhalation of oil mist. All used oils should be treated as dangerous waste and should be handled according to legal requirements.

Environmental properties comprise here both local and global environments. By local environment is meant the working environment with properties reducing, for example, oil mist or allergies when brought into contact with food and humans.

The global environment considers the impact if a lubricant is spilled on to the ground, but also considers the fact of oil depletion in the world. Thus, for example, biodegradability, toxicity and renewability are important factors.

These environmental properties will be covered below, together with marketing aspects. However, the history of lubricants that reduce the environmental impact will be covered first.

2.3.1 Environmentally Adapted Lubricants

Environmentally adapted lubricants (EALs) are based on biodegradable base fluids with a high degree of renewable content, such as ester base fluids together with low toxic, low bioaccumulative additive technology. The label on the container will immediately inform the customer of the reduced environmental impact.

Lubricants that reduce the environmental impact are generally technically as good as regular ones and fulfil the same requirements. Thorough testing of both the content and of the usage in the application is common and needed before launching the products.

Lubricants produced with the purpose of reducing the environmental impact have been on the market since the late 1980s. The history of these products shows how the focus changed from the base fluids to the full application within about a decade. The lubricants formulated to reduce the environmental impact were initially based on either natural or synthetic esters (refer to Chapter 3 for details). The first applications with an obvious need for these products were total loss applications, such as chainsaw oils and hydraulic oils used in mobile equipment in, for example, forestry.

Natural and synthetic esters have a different chemistry from regular mineral-based base fluids. Therefore, they interact differently with, for example, elastomer and bearing materials in the application. Consequently, there was a need to consider the total application and not only the lubricant itself. New products evolved with a stronger holistic view, including both base fluids and additives, in order to find a functioning solution for the application.

Some lubricants are used in the food industry. They can therefore not contain any toxic substances and they should also be tasteless and odourless. These are marketed as food grade. In addition, they should resist degradation from food products, and from chemicals and steam used when cleaning the production facilities.

Choosing lubricants based on, for example, white oils (refer to Chapter 3) instead of regular mineral base oils improves the working environment. Also, reducing the content of highly volatile components or base fluids can minimize oil mist.

The biodegradation of a lubricant is done by naturally occurring microorganisms. Water and carbon dioxide form when lubricants biodegrade. Biodegradability is tested on all base fluids used for formulating EALs. A 60% biodegradation within 10 days is the requirement for readily biodegradable lubricants.

Renewability is becoming increasingly important due to worldwide oil depletion. A source is considered renewable if it can be renewed within approximately 100 years.

Toxicity of base fluids and additives is measured on different biological organisms such as algae, daphnia and fish. Skin sensing tests are done on, for example, guinea pigs and rabbits. Tests are also performed to determine the lethal toxicity of both base fluids and additives. Base fluids are in general not toxic, while additives may display higher toxicity.

Bioaccumulation indicates the risk of accumulation within the food chain. Oil-soluble components have a greater risk of bioaccumulation. It may disturb reproduction and is therefore tested on both base fluids and additives. Base fluids in general do not bioaccumulate, while additives may bioaccumulate.

2.3.2   *Market Products with a Reduced Environmental Impact*

Customer awareness can be reached either top-down or bottom-up. By this is understood that legislation is required in some markets and that consumer influence is the driving force in other markets. Regardless of the driving force, ecolabels are a good indicator for the customer to indicate an environmental choice when purchasing the lubricant.

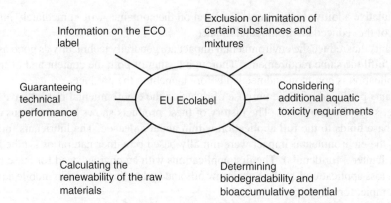

Figure 2.23 The contents of the EU Ecolabel serves as an example for parts to consider for lubricants developed to be more adapted to the environment

The aims of the ecolabels are to promote products that show a reduced impact on water and land in comparison to ordinary products. These products should contain a large fraction of biobased material. In the European Union, the EU Ecolabel can be awarded, for example, for hydraulic fluids, greases, total loss lubricants (e.g. chainsaw oils and stern tube lubricants), two-stroke oils, and industrial and marine gear oils. This is one way of pinpointing important aspects of environmental impact (see Figure 2.23).

A product having the ecolabel should not be formulated with substances exposing the user or the environment to different risks, as indicated by risk or hazard phrases shown on the label on the container. These hazard phrases can, for example, be *toxic if inhaled* or *hazardous to the ozone layer*. Some substances are known to be more hazardous than others and are therefore identified as *substances of very high concern* (i.e. SVHC) and should be avoided during formulation work.

Some lubricants are used at, for example, platforms or in ships where the exposure to aquatic life is considered to be high. In order to receive the EU Ecolabel, additional aquatic toxicity requirements must be met for the substances included in the additive. However, some substances are already known to be very toxic to aquatic life. They are therefore pinpointed and limited to be used in very small amounts only (i.e. quantities below 0.010 wt% of the final product).

The supplier of the product should prove biodegradability and bioaccumulative potential. Also, the renewable content in the product should be calculated and verified.

The products with a reduced environmental impact should meet minimum technical performance requirements. This technical performance is stated in ISO 15380 for hydraulic oils and according to DIN 51517 for gear oils (refer to Chapter 5 for more details).

The customer should be informed about the fact that the product has met the criteria of the ecolabel. This is communicated via the label on the container.

2.4 Summary of Analyses

Examples of standard methods for characterization of the lubricant properties described in the previous sections are given in Table 2.1.

Table 2.1 Standard methods for lubricant properties characterization

Method	Property	Description	Unit
ASTM D 445	Kinematic viscosity	KV_{40}	mm²/s
ASTM D 445	Kinematic viscosity	KV_{100}	mm²/s
ASTM D 5293	Dynamic viscosity	Cold cranking simulator (CCS)	mPa s
ASTM D 3829	Dynamic viscosity	Mini rotary viscosimeter (MRV)	mPa s
ASTM D 2983	Dynamic viscosity	Brookfield	mPa s
ASTM D 5481	Dynamic viscosity	High temperature high shear (HTHS)	mPa s
ASTM D 97	Pour point	Pour point	°C
ASTM D 4052	Density	ρ	kg/m³
ASTM D 3418	Specific heat capacity	C_p	J/kg, °C
ASTM D 5930	Thermal conductivity	λ	W/m, °C
CEC-L-40-A-93	Volatility	Noack	g/g (%)
ASTM D 92	Flash point	Cleveland open cup	°C
ASTM D 93	Flash point	Pensky Martin	°C
SS-ISO 6614	Filterability		ml/min
ASTM D 3427	Air release	Deaeration	min
ASTM D 892	Foaming		ml
ASTM D 1401	Demulsibility		ml
CEC-L-45-A-99	Shear stability	KRL (konische Rollenlager – tapered roller bearing)	%
ASTM D 665	Steel corrosion	Corrosion of iron	Pass/fail
ASTM D 130	Yellow metal corrosion	Copper corrosion	1A–3B
VDMA 24570	Soft metal corrosion	Linde	g
ASTM D 2272	Oxidation	RPVOT (rotating pressure vessel oxidation test)	min
ASTM D 6971	Oxidation	Ruler	%
ASTM D 943	Oxidation	TOST (turbine oxidation stability test)	hours
DIN 57370	Oxidation	Baader	hours
ASTM D 6186	Oxidation	PDSC (pressure differential scanning calorimeter)	%
ASTM D 2619	Hydrolysis	Coke bottle test	mg KOH/g
ASTM D 6304	Water content	Karl Fisher	g
ASTM D 95	Water content	Distillation	g
ASTM D 974	TAN	Colorimetric titration	mg KOH/g
ASTM D 664	TAN	Potentiometric titration	mg KOH/g
ASTM D 2896	TBN	Perchloric acid	mg KOH/g
ASTM D 4739	TBN	Hydrochloric acid	mg KOH/g
ASTM D 5381	Elemental analysis	XRF (X-ray florescence)	
ASTM D 7439	Elemental analysis	ICP (inductivity coupled plasma)	
ASTM D 7414	Molecular analysis	FTIR (Fourier transform infrared)	
OECD 301 B	Biodegradation		%
ASTM D 6866	Renewability	Theoretical calculation	%
ASTM D 6081	Additive toxicity	Exposure to, for example, daphnia and fish	mg

References

[1] Bird, R.B., Stewart, W.E. and Lightfoot, E.N. (1960) *Transport Phenomena*, Wiley International Edition, John Wiley & Sons.
[2] Briant, J., Denis, J. and Parc, G. (1989) *Rheological Properties of Lubricants*, Editions Technip, Institut Francais du Pétrole, France.
[3] Roelands, C.J.A. (1966) *Correlational Aspects for the Viscosity–Temperature–Pressure Relationship of Lubricating Oils*, Druck VRB, Gröningen.
[4] Larsson, R., Kassfeldt, E., Byheden, Å. and Norrby, T. (2001) Base fluid parameters for elastohydrodynamic lubrication and friction calculations and their influence on lubrication capability. *Journal of Synthetic Lubrication*, **18**(3), 183–198.
[5] Barus, C. (1893) *Note on the dependence of viscosity on pressure and temperature*, in Proceedings of the American Academy of Arts and Sciences, Boston, University Press, John Wilson and Son, **19**, 13.
[6] Totten, G.E., Shah, R.J. and Westbrook, S.R. (2003) *Fuels and Lubricants Handbook: Technology, Properties, Performance and Testing*, ASM International.
[7] Hunter, R.J. (1999) *Introduction to Modern Colloid Science*, Oxford Science Publications.
[8] Moore, W.J. (1983) *Basic Physical Chemistry*, Prentice-Hall International Edition.
[9] Holmberg, K., Jönsson, B., Kronberg, B. and Lindman, B. (2003) *Surfactants and Polymers in Aqueous Solutions*, 2nd edn, John Wiley & Sons.
[10] Bishop, R.J. and Totten, G.E. (2000) *Managing Foam and Aeration in Hydraulic Fluids*, in Proceedings of the Practicing Oil Analysis 2000 Conference, Noria Corp., Tulsa.
[11] Atkins, P and de Paula, J. (2009) *Atkins Physical Chemistry*, Oxford University Press.

3
Base Fluids

The terminology base fluid and base oil is commonly used. Here, base fluid will be used as a general term. The term base oil is used to describe base fluids originating from crude oil. Available base fluids have different properties depending on their chemical structures. Therefore, it will be important to cover some general hydrocarbon chemistry for a better understanding of base fluid properties.

3.1 General Hydrocarbon Chemistry

Hydrocarbons consist of carbon atoms and hydrogen atoms (see Figure 3.1). Carbon prefers to be bonded to four atoms, in this case four hydrogen atoms. Such hydrocarbons are denoted saturated. When carbon is bonded to less than four atoms (i.e. three or two atoms[1]) double or triple bonds arise and the hydrocarbon is called unsaturated [1].

The hydrocarbon molecular structures can be depicted in different ways. One explicit way is to show the carbon atom positions by a C and the hydrogen atom positions by an H connected by a chemical bond visualized by a line. However, this way becomes complicated for hydrocarbons due to the large molecular structures, and shorter ways have emerged, for example the carbon atoms and the hydrogen atoms are just imagined by lines in a molecular skeleton. For an even shorter version the letter R is used to describe a hydrocarbon chain and the end is depicted in detail, for example when the hydrocarbon chain ends with a hydrogen atom, the structure is visualized as RH. The chemical formula for such molecule is C_nH_{2n+2}, where n gives the number of C atoms.

The hydrocarbon structures present in base oils are paraffins, naphthenes and aromatics (see Figure 3.2). The hydrocarbon structures occur as *straight, linear hydrocarbons* as visualized in Figure 3.1, called normal paraffins (or n-paraffins), *with side chains* to the linear hydrocarbon chain (i.e. branched paraffins or iso-paraffins), *with at least one double bond* (i.e. unsaturated hydrocarbons or olefins), *in ring structures* (i.e. cycloparaffins or naphthenes) and aromatics.

---

[1] Carbon is bonded to three atoms in double bonds and to two atoms in the triple bond.

Lubricants: Introduction to Properties and Performance, First Edition.
Marika Torbacke, Åsa Kassman Rudolphi and Elisabet Kassfeldt.
© 2014 John Wiley & Sons, Ltd. Published 2014 by John Wiley & Sons, Ltd.

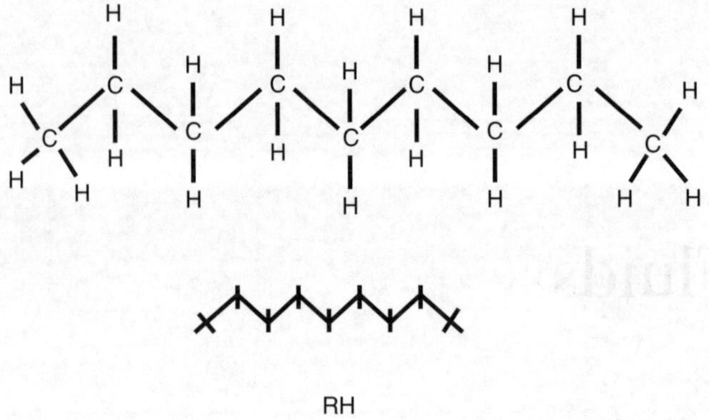

Figure 3.1 Three ways of describing a hydrocarbon. Top: the bonds of the carbon and hydrogen atoms are shown. Middle: the general structure is shown. Bottom: short description of where R is used for a hydrocarbon chain and H indicates one hydrogen at the end

The hydrocarbon chemistry will affect the properties of the base fluid and the formulated lubricant (see Table 3.1). The properties of base fluids depend not only on their chemical structure but also on their molecular weight distribution. In general, a narrow molecular distribution means more predictable properties.

Base fluid chemistry influences the performance of the lubricated contact. Here volatility, solvency, polarity (or surface activity), oxidation stability, viscosity and viscosity index will be discussed.

Some base fluids are composed of low weight molecular hydrocarbon structures (i.e. low viscosity base fluids). Low molecular weight hydrocarbons evaporate at elevated temperatures, giving an increased viscosity of the remaining base fluid. Consequently, *volatility* has an effect on lubricant consumption at high operating temperatures, particularly if the lubricant is based

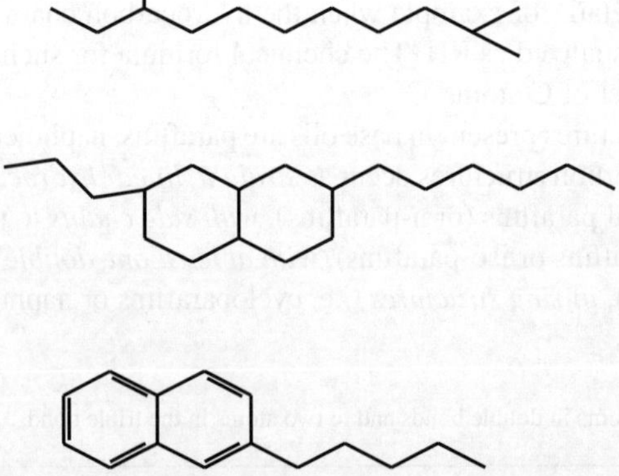

Figure 3.2 Different hydrocarbon structures in the crude oil: iso-paraffins (top), naphthenes (middle) and aromatics (bottom)

Table 3.1 Some properties of different hydrocarbon chemical structures

Property	Paraffinic	Naphthenic	Aromatic
Viscosity index	High	Low	Low
Density	Low	Low	High
Pour point	High	Low	Low
Volatility	Low	Medium	High
Flash point	High	Low	Low/medium
Oxidation stability	High	High	Low
Thermal stability	Low	Low/medium	High
Toxicity	Low	Low	Medium[a]
Elastomer compatibility	Shrink	Swell	Swell

[a]Carcinogenic

on low viscosity base fluids. Evaporation of low viscosity base fluids may cause deposit formation, for instance on cylinder walls in an engine, due to a change in solubility resulting in precipitation of low soluble compounds.

The base fluid chemistry will affect the ability to dissolve additives as well as contaminants. Thus, different base fluids have different *solvency* (i.e. the ability to dissolve solids, liquids or gases) depending on their chemical composition. Ring structures in the hydrocarbon will interact better than linear hydrocarbons with additive chemistries. Thus, naphthenes have a higher solvency than n-paraffins. The solvency will affect the stability of the formulated lubricant. In general, the higher the base fluid solvency the more stable is the final lubricant.

The *polarity* (or surface activity) determines both the ability to interact with additives and also the ability to interact with available surfaces. This implies that the base fluid polarity will affect foaming, air release, demulsification as well as adsorption to the lubricated contact surface. Therefore, the base fluids have to be selected with care for the final lubricant to function well in the application.

Hydrocarbons will oxidize at elevated temperatures. Thus, the *oxidation stability* of the base fluid affects the performance of the lubricant at higher application temperatures. Oxidation takes place when oxygen reacts with the hydrocarbon. However, oxygen is selective for certain hydrocarbon structures and parts of a hydrocarbon chain. The reactivity of different chemical structures is (in falling reactivity order): triple bonds, double bonds, ring structures, linear structures and branched structures. Consequently, double bonds are more easily attacked by oxygen than single bonds and n-paraffins are more prone to oxidize than iso-paraffins.

Hydrocarbons occur in different chain lengths, giving different base fluid *viscosities*. Short hydrocarbons result in low viscosity base fluids and long hydrocarbons in high viscosity base fluids. The chemical structure shows different temperature dependence and thereby gives the base fluids different *viscosity indices*. A base fluid with a high viscosity index can operate in a wide temperature range. Branching and ring structures yield improved cold flow properties whereas ring structures actually inhibit wax formation. Consequently, naphthenic and aromatic structures evince better cold flow properties than paraffinic structures [2].

3.2 Base Fluid Categorization

Base fluids are categorized for practical purposes offering a common terminology. The largest volumes of base fluids are used in lubricants for automotive applications. Engine testing for engine oils is intensive and costly. Original equipment manufacturers (OEMs) have very strict requirements for the lubricants in order to guarantee the functioning of the vehicle. Consequently, the global base fluid offering is to a large extent dependent on the approvals made by OEMs for the automotive segment. Categorization offers guidelines, for example, for approved base oil interchange in automotive applications.

In the automotive context it is convenient to divide the base fluids according to groups I, II, III, IV and V (see Table 3.2) [3]. It is apparent that this classification of base fluids is mainly for the automotive segment since group V contains all base fluids that cannot be fitted into another group. As a matter of fact, these base fluids are not commonly used in engine oil applications.

For industrial lubricants and/or environmentally adapted lubricants (EALs) the description in Figure 3.3 is more useful. This diagram shows base fluids originating either from natural gas or coal, crude oil or from renewable raw materials (i.e. of plant and animal origin). Natural gas or coal is the raw material for producing gas-to-liquid (GTL) base fluids. Crude oil gives rise to several base oils (paraffinic and naphthenic base oils, white oils, very high viscosity index (VHVI) base oils and polyalphaolefins (PAOs)).

Starting with renewable raw materials yields vegetable oils and synthetic esters that are more readily biodegradable than base oils originating from crude oil. Together with renewability, biodegradability is an important property for EALs. Thus, by moving to the right in Figure 3.3 the biodegradability and the renewability increases, and consequently so does the final lubricant as well.

Table 3.3 describes the categories of base fluids, combining Table 3.2 and Figure 3.3. The common names are also given. It is apparent that the two ways to categorize base fluids supplement each other. Each group I to III includes several base oils with different origins. Group V also includes a very broad range of base fluids where other descriptive ways are more informative.

The market needs for base fluids depend on availability, performance requirements and the offered price. Currently group I base oils are dominant (see Figure 3.4). However, group II base oil volumes are increasing. Also, group III base fluid volumes are increasing due to increased

Table 3.2 Classification and properties of the base fluids, as presented by API[a] and ATIEL[b] [3]

Group	Description
Group I	Saturated hydrocarbons < 90%, sulfur > 0.03%, $80 \leq VI \leq 120$
Group II	Saturated hydrocarbons $\geq$ 90%, sulfur $\leq$ 0.03%, $80 \leq VI < 120$
Group III	Saturated hydrocarbons $\geq$ 90%, sulfur $\leq$ 0.03%, $VI \geq 120$
Group IV	PAOs
Group V	All others, but groups I, II, III and IV: e.g. naphthenics, synthetic and natural esters

[a] American Petroleum Institute
[b] Association Technique de l'Industrie Européenne des Lubrifiants

Base Fluids 49

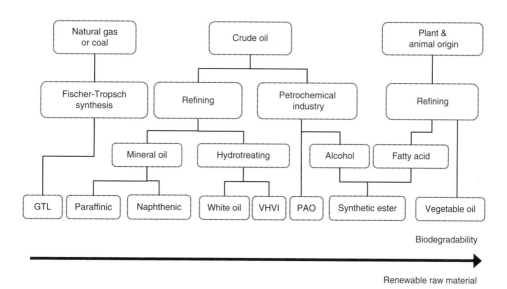

Figure 3.3 The origin of different base fluids (For a colour version of this figure, see the colour plate section)

Table 3.3 Origin and description of different base fluids

API group	Description	Origin	Common name
I	Paraffinic	Crude oil	Mineral oil
I	Re-refined	Used oils	Re-refined
II	Paraffinic	Crude oil	Mineral oil
II	Re-refined	Used oils	Re-refined
II	White oils	Crude oil	White oil
III	VHVIs	Crude oil	Synthetic fluid
III	GTLs	Natural gas (or coal)	GTLs
IV	PAOs	Crude oil	Synthetic fluid
V	Naphthenic	Crude oil	Mineral oil
V	Vegetable oils	Vegetable oils	Vegetable oil
V	Synthetic esters	Vegetable oils and crude oil	Ester

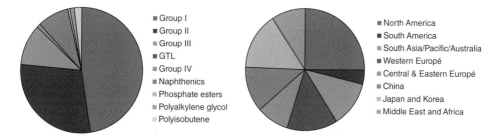

Figure 3.4 The relative production of different base fluids according to figures from refineries across the world (left). The production of base fluids from different parts of the world (right)

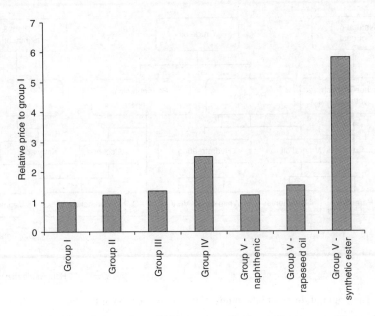

Figure 3.5 The relative prices of different base fluids relative to group I base oils

technical demands, such as increasing emission demands. Most base fluids are refined in North America, Western Europe, China, Japan and Korea.

The relative prices of base fluids are given in Figure 3.5. Currently, group I base oils are the lowest priced base oils, followed by group II base oils, making them attractive alternatives.

3.3 The Refining Process of Crude Oils

Crude oils are hydrocarbon molecules with small amounts of sulfur, oxygen, nitrogen, inorganic salts and metals. The quality of crude oils varies from oil well to oil well. In general the quality is in the ranges given in Table 3.4 [4]. In this section the refining process will be described.

Table 3.4 Typical composition of crude oils

Component	Wt%
Carbon	83–87%
Hydrogen	11–14%
Sulfur	0–3%
Nitrogen	0–1%
Oxygen	0–0.5%
Metals	0–0.2%

3.3.1 The Refining Process

As has been mentioned earlier, crude oil is the basis for refining or synthesizing mineral base oils with a content of primarily paraffinic and naphthenic structures. The crude oil needs to be processed in order to retrieve base oils suitable for lubricant production. A schematic refining process scheme is shown in Figure 3.6. The refining process separates the crude oil into its constituent molecules. As a final step the base oil is reacted with hydrogen.

Refining is a balance between removing undesirables while keeping desirables. Desirables can be sulfur compounds, which are natural antioxidants. The purpose of the refining process is to obtain base oils with an even and defined quality that are possible to use in lubricant production.

The refining process starts with distillation at atmospheric pressure. The crude oil stream is heated and there will be a temperature gradient across the distillation tower, where the top of the tower will have a lower temperature and the bottom of the tower will have a higher temperature. The crude oil molecules are separated according to their difference in volatility. The more volatile molecules, that is the low weight molecules, will evaporate and move to the top and the less volatile molecules, that is the high weight molecules, will remain in the liquid phase and move downwards to the bottom (see Figure 3.6) [5–7].

The low weight molecules (i.e. light fractions) will continuously be removed from the top of the tower. These are gases, gasoline, naphtha, kerosene and light gas oil. They are further distilled and separated in another unit to be prepared for the fuels market. Fuels will not be further covered in this section.

The high weight molecules are fed to a vacuum distillation tower. Distillation under vacuum is carried out to allow for separation at lower temperatures. This is required when the feed stream consists of large molecular structures with low volatility. In the same manner as for the distillation at atmospheric pressure, evaporation of the lower weight molecules takes place.

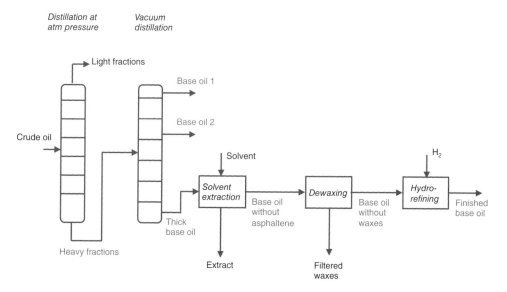

Figure 3.6 A simplified picture of the refining process of crude oil

Table 3.5 The influence of crude oil processing on some base oil properties [4]

Process/property	Deasphalting	Dewaxing	Hydrorefining
Viscosity	Decrease	Increase	No change
Viscosity index	Increase	Decrease	No change
Density	Decrease	Increase	Decrease
Pour point	Increase	Decrease	Increase
Flash point	No change	No change	No change
Oxidation stability	Improve	No change	Improve
Additive response	Improve	No change	Improve
Colour	Improve	Depends	Improve

In this case the low weight molecules consist of base oils with low viscosity. It is possible to remove fractions at different heights of the distillation tower according to the desired viscosity. At the bottom of the vacuum distillation tower thick base oil is collected [4, 6, 7].

The thick base oil contains undesirable components such as asphaltenes and waxes. Asphaltenes can be removed by solvent extraction. This step will improve the oxidation stability and reduce formation of sludge. Solvent extraction is a liquid–liquid extraction where separation is possible due to the relative difference in solubility in two different immiscible liquids. Thus, a solvent is added to the thick base oil and the asphaltene is extracted to the solvent.

The asphaltene-free base oil still contains waxes. These waxes crystallize at relatively high temperature, giving the final base oil a high pour point, which is undesirable. Thus, the properties will be greatly improved by removing the waxes either by cooling and allowing the waxes to crystallize (or by solvent extraction). Cooling allows the separation due to differences in solubility with temperature between the waxes and the remaining base oil. Thus, the temperature is lowered and the base oil will be dewaxed [8].

Final hydrorefining[2] is done to meet colour and oxidation stability requirements. It involves treating the base oil with hydrogen and converting unsaturated hydrocarbons to saturated hydrocarbons. Also, sulfur, nitrogen and impurities are removed. The sulfur and the nitrogen content depend on the crude source. High amounts of sulfur are commonly found in aromatic base oils. However, they have limited use as they are carcinogenic. Sulfur compounds act as natural antioxidants. Despite this fact it is sometimes desirable to remove sulfur. This can be the case for lubricants used in engine oils where the amount of sulfur should be kept at a minimum to ensure emission levels. Nitrogen increases oxidation of the base oil and is therefore removed. Hydrorefining is particularly desired for white oils used in the pharmaceutical and the food industry [4].

3.3.2 Influence of the Refining Process on the Oil Properties

Crude oil processing affects the base oil properties (see Table 3.5). The distillation steps separate the total crude into hydrocarbons with different chain lengths, yielding different base

---

[2] Sometimes hydrofinishing is used. Hydrorefining is a more severe process involving cracking (or the breaking of molecular structures) than the hydrofinishing process.

oils. Therefore, the properties cannot be described before and after distillation. All the following separation methods remove chemical compounds physically from the base oil without affecting the hydrocarbon structures and the property changes can therefore be described.

The refining process results in several base oils with a variety of properties. The different base oils used in lubricants will be explored below and described according to their general appearance, molecular structure, significant properties and usage.

3.4 Base Fluids Originating from Crude Oil

Base oils will be described following the logical order of Figure 3.3. Those originating from crude oil will be divided into paraffinic base oils, naphthenic base oils, white oil and VHVI base oils, PAOs and re-refined base oils. GTL oils will be covered here due to their chemical similarity to crude-based base oils. The general properties and the chemistry will be described together with their fields of application. Also, a link between base oil chemistry and properties will be given.

Hydrocarbons used in lubricants have three basic chemical structures: paraffinic, naphthenic and aromatic. The distribution between the chemical structures varies between different base oils, giving them their unique properties. The structures have 20 or more carbons in the hydrocarbon chain. Base oils with fewer carbons usually have too low viscosity and correspondingly too high volatility to be useful in lubricants [4, 5, 9].

3.4.1 Paraffinic Base Oils

Paraffinic base oils consist of 45–60% paraffinic structures. They are commonly light yellow due to a small amount of aromatics. This colour becomes more intensely yellow and slightly fluorescent as the viscosity increases. One reason for the colour to deepen with viscosity is that the thickest base oils are made from paraffinic and naphthenic residues, that is taken from the bottom of the distillation unit containing more aromatic compounds. These highest viscosity paraffinic base oils are commonly referred to as bright stocks.

Paraffinic base oils have a favourable price and are used extensively in most lubricants. They have good lubricating properties and additive response. They can be used in temperatures up to 200 °C in the absence of oxygen and up to 150 °C in the presence of oxygen. They have relatively good low temperature properties, which can be further enhanced with additives. In addition, paraffinic base oils have inherently high viscosity indices. They are used in lubrication regimes from full film to boundary lubrication. The largest volumes of paraffinic base oils are used in engine oils, automotive transmission oils and industrial lubricants [4].

3.4.2 Naphthenic Base Oils

Naphthenic base oils consist of 65–75% naphthenic structures and 25–35% paraffinic structures. They are usually light yellow due to low levels of aromatics. The naphthenic structures allow for good solubility of additives. Naphthenic base oils are sometimes used in lubricants of other base oils in order to improve the solubility of the additives. They show good low temperature properties, low viscosity index and good solubility properties. However, the poor viscosity indices (VIs) do not make them suitable for high temperature applications [4].

Naphthenic base oils are less common than paraffinic base oils in oil wells. Only 10% of petroleum base oils are naphthenic base oils. They are naturally wax free or low in wax content. Therefore, they have inherently excellent pour point properties. They are fairly low priced and are mainly used where their low pour point properties or high solvency is desired. Such applications involve hydraulic oils, turbine oils and metal-working applications [6].

3.4.3 White Oils

White oils result from the finishing step of the crude oil processing. They consist of a mixture of paraffinic and naphthenic structures. However, these oils have very low levels of sulfur, nitrogen and aromatics. They are colourless due to the lack of aromatics. The main use for white oils is in the pharmaceutical, food and cosmetic industries, where the legal requirements of cleanliness are high. In some cases lubricants are formulated with white oils for work-environment reasons. Also, white oils are used in, for instance, the textile industry, where the requirements for nonstaining are high. White oils are relatively expensive and they are therefore only used in special applications [4].

3.4.4 Very High Viscosity Index Base Oils

Very high viscosity index base oils (VHVIs) are hydroisomerized, implying that the molecular structures are rearranged in a predetermined way while keeping the molecular weight constant. They are further treated with hydrogen to remove double bonds and to break up ring structures in the hydrocarbons. VHVI base oils consist of mainly paraffinic structures, even though naphthenic structures occur. They have low levels of sulfur and nitrogen. As the name indicates, they have very high viscosity indices, but they have also low volatility, good oxidation stability and improved cold flow properties compared to group I and II base oils.

VHVI base oils are marketed as synthetic base oils, indicating that the chemical structure is tailored. This commonly gives a narrower molecular weight distribution with more predictable base fluid properties. They are used in, for example, engine oils where new emission standards require high quality base oils [2, 4].

3.4.5 Polyalphaolefins

Polyalphaolefins (PAOs) are synthesized from linear paraffins (1-decene molecules). It is the only 'true' synthetic base fluid originating from crude oil (or the petrochemical industry). The synthesis produces a base fluid of well-defined molecular structure with a very narrow molecular weight distribution. Examples of the molecular structures of different PAOs are shown in Figure 3.7. The narrow and well-defined molecular structure of the PAOs implies that they have both good high temperature properties and good low temperature properties. When no low weight molecular structures are present the volatility is improved. In addition, the PAO structure is fully saturated (i.e. no double bonds), which also improves the high temperature properties. Also, when there are no high weight molecular components present, the cold flow properties are improved.

Base Fluids

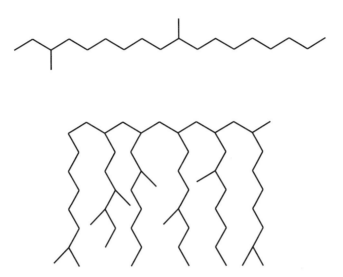

Figure 3.7 The molecular structures of PAO 2 (top) and PAO 6 (bottom)

The ability to dissolve additives is low for PAOs due to the well-defined molecular structure without double bonds and aromatic structures. Therefore, esters or naphthenic base oils are commonly added to improve the additive solvency. Also, PAOs are known to shrink elastomer materials.

PAOs show superior performance compared to mineral base oils, but are more expensive. The viscosity index is high, the volatility is low and the oxidation stability is good. The lower viscosity PAOs are shear stable and fully compatible with mineral oils. They are used in transmission oils and in 0W-xx engine oils because of their excellent low temperature properties [4, 10].

3.4.6 Gas-to-Liquid Base Fluids

Gas-to-liquid (GTL) base fluids have mainly been produced when there has been a limit of conventional mineral oil products. GTLs are considered too expensive today due to the high capital investment needed. Nevertheless, there are a few GTL production units. GTL base fluids are believed to be one of the base fluids of the future[3] when the hydrocarbon depletion reaches a critical stage and the price of conventional petroleum products increases.

GTL base fluids can be produced from natural gas or coal by a Fisher–Tropsch (F-T) process. The process to convert natural gas to liquids can be divided into three process steps: synthetic gas (syngas) generation, syngas conversion and hydroprocessing. The syngas generation process is expressed as

$$2\,CH_4 + O_2 \rightarrow 2\,CO + 4\,H_2 \tag{3.1}$$

---

[3] The GTL process is mainly utilized to produce fuels. Longer molecular structures will be used for producing GTL base fluids.

The syngas is the feed material to the F-T process where the syngas is converted to carbon–carbon bonds

$$n\text{CO} + 2n\,\text{H}_2 \rightarrow -(\text{CH}_2)_n\text{-} + n(\text{H}_2\text{O}) \qquad (3.2)$$

The F-T products are comprised of linear paraffinic structures of a wide molecular weight range. The F-T products are unique with zero content of sulfur, nitrogen, aromatics, olefins and metals. GTLbase fluids are similar to VHVI base oils and PAOs with a high VI, low volatility and good cold flow properties. They are classified as group III base fluids and marketed as synthetic base fluids [11–13].

3.4.7 Re-Refined Base Oils

Any mineral-based lubricant or industrial oil that is not suitable to be used should be re-refined rather than burnt, according to the EU's Waste Directive from 2008[4] [14]. The re-refined base oil production was estimated at 500 000 tons or 8% of the lubricant consumption in 2009.

Lubricants age when being used. This implies that additives are being used and that base oils degrade. In addition, the lubricant may have become contaminated with water, dissolved contaminants or particles. After a while, the lubricant is not suitable in the application and needs to be replaced. However, the used lubricant can be re-refined by purification, removal of water, particles and soluble contaminants. The re-refining process results in high quality base oils.

The re-refining process is similar to refining crude oil. The used lubricants are first distilled under atmospheric pressure to remove water and volatile components such as fuel. The next step is to distil the heavy fractions into suitable fractions. The final purification step removes the remaining undesirables such as particles or dissolved contaminants with solvent extraction. The solvents are separated from the base oils.

All re-refiners in Europe produce group I and group II base oils. Re-refined base oils are of high quality with low volatility and pour points like other group I base oils. Surface active components may be difficult to remove, which may influence foaming, deaeration and demulsification properties [14, 15].

3.5 Base Fluids Originating from Renewable Raw Materials

Vegetable oils or animal fats have been used for thousands of years as lubricants. As mineral-based lubricants have a better cost performance ratio, vegetable oils or animal fats have been less attractive alternatives. Vegetable oils have been used extensively during times of mineral base oil shortage [16]. Now we are facing a hydrocarbon depletion reality that will force us to look at alternatives to mineral-based lubricants making environmentally adapted lubricants (EALs) more competitive. The demand for EALs has been particularly strong in Sweden, Germany and the Alpine region [16–19].

---

[4] In 2009, 5.7 million tons of lubricants were sold in the EU. Half of this volume was estimated to be collectable. Nevertheless, only 2 million tons were collected.

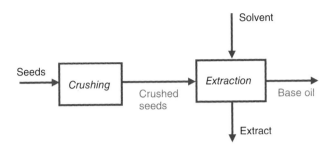

Figure 3.8 General picture of oil extraction from a seed

3.5.1 Vegetable Oils (Natural Esters)

Vegetable oils consist mainly of triglycerides (i.e. glycerol with three fatty acids). The most commonly used vegetable oils for lubricants are high oleic rapeseed oils, sunflower oils, soybean oils and castor oils. Rapeseeds have an oil content of 40–45%, sunflower seeds contain 40% oil and soybeans contain 20% oil. Currently, most vegetable oils from these sources are converted to fatty acid methyl ester, that is the biopart in biodiesel fuels. Soybeans are mostly grown in the United States and vegetable oils made from soybeans are consequently produced and mainly used in the United States. Rapeseeds, sunflower seeds and soybeans contain mostly oleic acid and linoleic acid, although the proportions differ. Oleic acid has 18 carbons and one double bond. Linoleic acid has 18 carbons and two double bonds.

The seeds need to be crushed to retrieve the oil. This oil is further refined by solvent extraction to remove undesirables and prolong the life of the vegetable oil (see Figure 3.8).

Vegetable oils have excellent lubricating properties, inherent high VIs, high flash points, are biodegradable and renewable, and are in general low toxic. However, the oxidation stability is low due to the occurrence of double bonds (i.e. unsaturated). If untreated, they will quickly oxidize, entailing increased viscosity, and polymerize, producing a plastic-like surface. Chemical modifications of vegetable oils with hydrogen improve their oxidation stability.

Vegetable oils have limited cold flow properties. The pour point of a vegetable oil-based lubricant is improved by adding pour point depressants or adding another base fluid with superior pour point properties. In the case of EALs, this other base fluid is usually a synthetic ester. Lubricants made of vegetable oils are mostly used in total loss applications such as hydraulic oils and chainsaw oils. For instance, methanol engines are lubricated with vegetable oils (i.e. castor oils).

3.5.2 Synthetic Esters

Natural and synthetic esters are structurally similar. However, synthetic esters are manufactured and tailored to a specific structure. Synthetic esters are produced by reacting an alcohol with a fatty acid where R_1 and R_2 are hydrocarbons with different number of carbons

$$R_1 - OH + R_2 - COOH \leftrightarrow R_1 - COO - R_2 + H_2O \qquad (3.3)$$

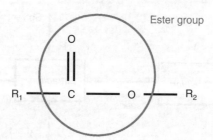

Figure 3.9 A schematic of a monoester, where COO is the ester group (For a colour version of this figure, see the colour plate section)

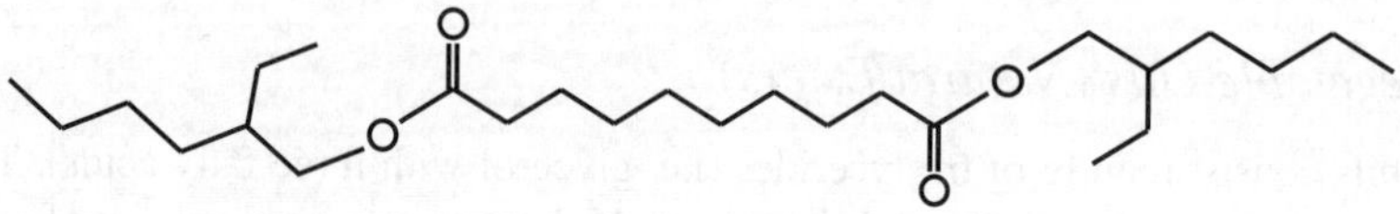

Figure 3.10 The molecular structure of a diester

The alcohol is usually of petrochemical origin while the fatty acid is available from natural vegetable oils, animal oils or fats. There is a wide variety of both fatty acids and alcohols. Thus, there is a wide range of esters available. For instance, there are monoesters, diesters and polyolesters. Monoesters have one ester group, diesters have two ester groups and polyolesters have several ester groups (usually three) [20].

Monoesters (see Figure 3.9) generally have low viscosity and high volatility. They are mostly used in metal-working applications. *Diesters* (see Figure 3.10) have good VIs and good low temperature properties. However, their use is limited owing to fairly low viscosities and high seal swell. Sometimes diesters are added to PAO blends to provide controlled increased seal swell.

Polyolesters (see Figure 3.11) have similar properties to diesters, but with a wider range of viscosities available. They are often hydrolytically, and thermally, more stable than diesters. Polyolesters are used in EALs where there is a need for a biodegradable, renewable and low

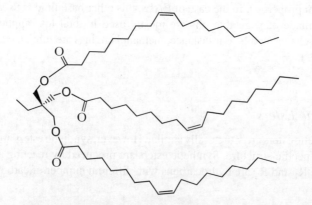

Figure 3.11 The molecular structure of a polyolester

toxic base fluid. Also, polyolesters are added to mineral base oils in order to improve the solubility of additives.

3.6 Nonconventional Synthetic Base Fluids

Nonconventional synthetic base fluids comprise, for example, phosphate esters, polyalkylene glycols, polyisobutenes and silicone oils. They are generally expensive and are used when other base fluids cannot perform due to, for example, very high loads and very high temperatures. Not all of them are miscible with other base fluids. The nonconventional base fluids mentioned above will be described shortly [21].

Phosphate esters have high flash points and low vapour pressures, making them suitable in high temperature applications. They are fire resistant and therefore commonly used due to their relative competitive pricing, being a synthetic base fluid. They show good hydrolytic and oxidative stability in addition to good thermal stability. They also show good cold flow behaviour even though their viscosity temperature behaviour is considered poor. Their chemistry is based on carbon, hydrogen, oxygen and phosphorus. They are not miscible with mineral base oils.

Polyalkylene glycols have a very high oxygen content, making them unique in many ways. They can be used where other lubricants would result in sludge formation. They can be tailored, allowing them to be either oil or water soluble. Their chemical structure is based on carbon, hydrogen and oxygen. This allows them to show extraordinary high thermal and oxidative stability in combination with low volatility, which makes them suitable in applications with high temperatures due to being fire resistant. They have good cold flow behaviour. Unfortunately, they are not miscible with mineral base oils. At very high temperatures they may form toxic decomposition vapours. Their cost is very high.

Polyisobutenes are hydrocarbon polymers used in, for example, lubricants. For instance, they are used instead of very high viscosity mineral base oils due to their special properties, such as low deposit formation, low toxicity, good thickening and high shear stability. In addition, they are the basis in detergent additives added to lubricants and are also used as a smoke inhibitor in two-stroke engine fuels. They are commonly described by their viscosity and flash point. They are miscible in mineral base oils. However, they have moderate oxidative stability, high volatility and moderate cold flow behaviour.

Silicone oils are used as, for example, fluids and elastomers. In lubrication, they are used in both metal and nonmetal contacts. They show good viscosity temperature behaviour, high oxidative and thermal stability, and excellent cold flow behaviour. They have low volatility and correspondingly high flash points. They have low surface tension with a good wetting capacity. They have a low load-carrying capacity, which cannot be improved with additives. Therefore, they are poor at mixed and boundary lubrication. Their chemistry is based on carbon, hydrogen, oxygen and silicone. They are immiscible with mineral base oils as well as water. Their cost is very high.

3.7 Properties of Base Fluids

Some properties of mineral base fluids are summarized in Table 3.6 and of vegetable oils and esters in Table 3.7.

Table 3.6 The properties of mineral base fluids

Base Fluid	Name	KV_{40} (mm²/s)	KV_{100} (mm²/s)	VI (—)	Density (kg/m³)	Volatility (%)	Flash point (°C)	Pour point (°C)
Group I: Paraffinic	SN 100	20	4.0	95	860	30	194	−18
Group I: Paraffinic	SN 150	30	5.3	100	870	16	210	−12
Group I: Paraffinic	SN 600	112	12.0	95	880	—	246	−6
Group I: Paraffinic	2500 BS	450	31	95	—	—	294	−6
Group I: Re-refined	Re-refined 4	12.8	3.1	105	850	—	—	−15
Group I: Re-refined	Re-refined 5	29.5	5.4	115	854	9	—	−12
Group I: Re-refined	Re-refined 10	59	8.5	117	858	3	—	−9
Group II	Group II 4	20.4	4.1	102	—	26	210	−13
Group II	Group II 6	41.5	6.4	103	—	11	226	−12
Group II	Group II 10	113	12.4	101	—	2	247	−13
Group III	VHVI 4	20.5	4.3	121	831	14.4	224	−18
Group III	VHVI 5	26	5.1	126	835	8.4	240	−15
Group III	VHVI 6	34	6	128	838	5.7	240	−15
Group III	VHVI 8	50	8	128	843	3.5	260	−15
Group IV	PAO 4	17	3.9	120	820	14	200	−65
Group IV	PAO 6	31	5.9	135	825	7	215	−60
Group IV	PAO 8	48	7.9	135	825	4	220	−55
Group IV	PAO 10	65.5	9.7	129	840	4	250	−45
Group V: Naphthenic	Group V 4	22	3.7	2	901	—	178	−42
Group V: Naphthenic	Group V 5	30	4.5	24	896	—	184	−39
Group V: Naphthenic	Group V 10	104	8.8	29	906	—	216	−27

SN X = solvent neutral (*solvent* from solvent extracted, *neutral* from neutralization after acid washing and X is the viscosity measured at 100 °F in the unit Saybolt seconds)
BS = bright stock (bright is used because of the fluorescence of the heavy aromatics)
Group III and group IV base fluids are numerated after the viscosity at 100 °C
The re-refined, group II and group V base fluids have been named in the same manner as the group III and the group IV base fluids in this table

Table 3.7 The properties of vegetable oils and esters. Values have been approximated for a range of monoesters, diesters and polyesters

Base fluid	KV_{40} (mm^2/s)	KV_{100} (mm^2/s)	VI (–)	Density (kg/m^3)	Flash point (%)	Pour point (°C)	Iodine number (°C)
Rapeseed oil	40	8.4	193	920	330	−20	110–126
Sunflower oil	40	—	—	—	265	—	—
Monoester	5–10	2–3	150–200	870	200	−21	80
Diester	10–40	3–8	160	900	250	−50	0.5
Polyolester (saturated)	15–70	3–8	150	950	275	−40	—
Polyolester (unsaturated)	20–90	7–12	200	920	300	−40	40–90

References

[1] Solomons, T.W.G. (1996) *Organic Chemistry*, 6th edn, John Wiley & Sons.
[2] Phillips, R.A. (1999) Highly refined mineral oils, in *Synthetic Lubricants and High-Performance Functional Fluids*, 2nd edn (eds L.R. Rudnick and R.L. Shubkin), Marcel Dekker Inc.
[3] ATIEL (2003) *Base Stock Quality Assurance and Interchange Guidelines*, Appendix B.
[4] Sequeira, A. Jr (1994) *Lubricant Base Oil and Wax Processing*, Marcel Dekker Inc.
[5] Casserly, E.W. and Venier, C.G. (1999) Cycloaliphatics, in *Synthetic Lubricants and High-Performance Functional Fluids*, 2nd edn (eds L.R. Rudnick and R.L. Shubkin), Marcel Dekker Inc.
[6] Coulson, J.M., Richardson, J.F., Backhurst, J.R. and Harker, J.H. (1999) *Chemical Engineering Volume 1: Fluid Flow, Heat Transfer and Mass Transfer*, Chemical Engineering Series, Butterworth-Heinemann.
[7] Perry, R.H. and Green, D. (1984) *Perry's Chemical Engineers' Handbook*, 6th edn, McGraw Hill International Edition.
[8] Mullin, J.W. (1993) *Crystallization*, 3rd edn, Butterworth-Heinemann Ltd, Oxford.
[9] Briant, J., Denis, J. and Parc, G. (1989) *Rheological Properties of Lubricants*, Editions Technip, Institut Francais du Petrole.
[10] Rudnick, L.R. and Shubkin, R.L. (1999) Poly-α-olefins, in *Synthetic Lubricants and High-Performance Functional Fluids*, 2nd edn (eds L.R. Rudnick and R.L. Shubkin), Marcel Dekker Inc.
[11] Vosloo, A.C. (2001) *Fischer–Tropsch: A Futuristic View, Fuel Processing Technology*, Elsevier Science B.V.
[12] Fleisch, R.H., Sills, R.A. and Briscoe, M.D. (2002) 2002 – Emergence of the gas-to-liquids industry: a review of global GTL developments. *Journal of Natural Gas Chemistry*, **11**, 1–14.
[13] Corke, M.J. (1998) GTL technologies focus on lowering costs. *Oil and Gas Journal*, 71–77.
[14] Council Directive 75/439/EEC of 16 June 1975 on the disposal of waste oils.
[15] DeMarco, N. (2009) EU Favors Rerefining Used Oil. Lube Report, March 11, 2009.
[16] Hartmann, C. (2009) The New European Waste Framework Directive and its impact on re-refining and base oils, in ICIS World Base Oils and Lubricants Conference.
[17] Honary, L.A.T. (2001) Biodegradable/biobased lubricants and greases, *Machinery Lubrication Magazine*.
[18] Torbacke, M., Norrby, T. and Kopp, M. (2008) Lubricant design for environment, in *Proceedings of NordTrib Conference*, Tampere, Finland.
[19] Norrby, R., Torbacke, M. and Kopp, M. (2002) Environmentally adapted lubricants in the Nordic marketplace – recent developments. *Industrial Lubrication and Tribology*, **54**(3), 109–116.
[20] Randles, S.J. (1999) Esters, in *Synthetic Lubricants and High-Performance Functional Fluids*, 2nd edn (eds L.R. Rudnick and R.L. Shubkin), Marcel Dekker Inc.
[21] Rudnick, L.R. and Shubkin, R.L. (eds) (1999) *Synthetic Lubricants and High-Performance Functional Fluids*, 2nd edn, Marcel Dekker Inc.

4

Additives

The base fluid has film-forming properties and thereby reduces, for example, friction and wear. Thus, the base fluid alone will cover more or less all aspects that we demand from a lubricant. In spite of that, additives are added. They enhance the characteristics of the base fluid, lending the lubricant its final properties. They are mainly added to the base fluid to enhance the viscosity index and the pour point, the friction and wear properties under boundary and mixed lubrication, and the lubricant life. As a result, modern lubricants allow for increased temperatures and increased loads in applications. This chapter will give a basic understanding of fundamental concepts and processes before exploring different additives used in lubricants.

4.1 Fundamental Concepts and Processes

This section starts with some general chemistry, that is atoms, bonds, forces, reactions and chemical potential. It continues with surfaces, mass transfer and adsorption, and gives a basic general description of surface active molecules. The purpose is to update the reader on the chemical concepts and processes necessary to understand how the base fluids and the additives are acting in the lubricant [1].

4.1.1 Atoms and Reactions

Each element in the periodic system consists of characteristic atoms. Atoms have a nucleus, with neutrons (without a charge, i.e. neutral) and protons (with a positive charge), which is surrounded by electrons (having a negative charge). The simplest atom is the hydrogen atom, shown in Figure 4.1 according to Bohr's atomic model. In the centre is the positive nucleus (one proton), which is surrounded by an electron moving in a path at a certain distance from the nucleus.

The oxygen atom has eight protons and eight neutrons and eight electrons. The electrons circulate around the nucleus in paths at certain distances. These paths are referred to as electron shells. The inner electron shell can only carry two electrons. Thus, the other six electrons must circulate around the nucleus at a further distance. This outer shell could carry eight electrons.

Lubricants: Introduction to Properties and Performance, First Edition.
Marika Torbacke, Åsa Kassman Rudolphi and Elisabet Kassfeldt.
© 2014 John Wiley & Sons, Ltd. Published 2014 by John Wiley & Sons, Ltd.

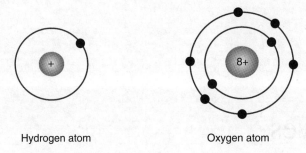

Figure 4.1 Bohr's atomic model of hydrogen and oxygen

Atoms prefer to have their outmost electron shell filled. If the outer shell is not filled initially, the atom may receive a filled outer shell by *attracting electrons* from neighbouring atoms, *donating electrons* to neighbouring atoms or by *sharing electrons* between neighbouring atoms. Molecules arise when chemical bonds are formed between atoms. The chemical bonds can be ionic,[1] where ions react, or covalent,[2] where electrons are shared between the atoms in the final molecule. A general description of a reaction can be visualized by reactants A and B reacting to form the product AB:

$$A + B \leftrightarrow AB \tag{4.1}$$

All reactions can go both to the right and to the left, which is shown by the arrow pointing in both directions. Equilibrium is reached when the reaction rate to the right equals the reaction rate to the left.

A reaction involving ionic bonds can be visualized by the formation of iron sulfide on iron surfaces in lubricants containing sulfur. An iron atom has two electrons in its outer shell, while a sulfur atom is lacking two electrons in its outer shell. When an iron atom is donating two electrons to a sulfur atom, iron and sulfur ions with filled outer shells are formed. These ions will be attracted by each other, resulting in ionic bonded iron sulfide (see Figure 4.2).

Reactions involving covalent bonds are represented by the reaction between carbon and hydrogen, forming a hydrocarbon (see Figure 4.3).

4.1.2 Intermolecular Forces

The lubricant consists of a large number of molecules, both base fluid and additive molecules, which interact with each other. The interaction can be strong or weak depending on the molecules involved. Attraction forces between nonpolar molecules, such as hydrocarbons, are weak because they are neutral with an even charge distribution. The weak forces between nonpolar base fluid molecules are referred to as *van der Waals forces*.

---

[1] Ionic bonding is formed by an attractive force between oppositely charged ions. Ion bonding occurs between atoms of large difference in electronegativity (ability to attract electrons).
[2] Covalent bonding occurs when two or more atoms of the same (or similar) electronegativity react. Electrons are not transferred; instead they are shared between the atoms.

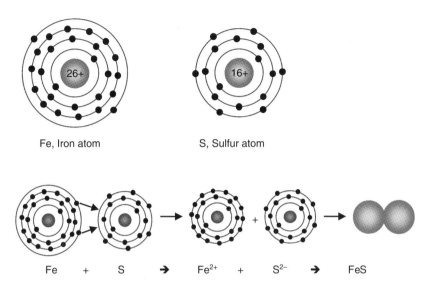

Figure 4.2 The models of iron and sulfur are shown at the top. The reaction between iron and sulfide forming iron sulfide is shown below

Polar molecules are often dipoles with a nonuniform charge distribution, that is a nonuniform distribution of the electrons. For example, in ester base fluids the electrons are distributed more towards the oxygen part of the molecule, turning the ester into a dipole. The molecules orientate themselves with the positive side pointing at the negative side of its neighbouring molecule. Another example is water (H_2O), with very strong *dipole–dipole attraction forces* between the polar water molecules.

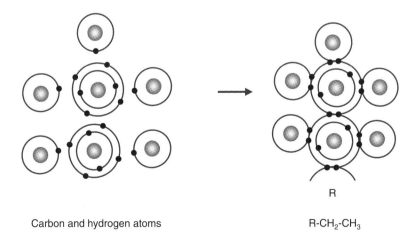

Figure 4.3 The reaction between carbon and hydrogen, forming a covalently bonded hydrocarbon

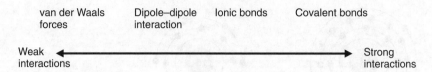

Figure 4.4 The relative strength of different interactions is visualized

Dipole–dipole interactions are stronger than van der Waals attractions. However, the magnitude of intermolecular forces is much lower than the strength of chemical bonds, such as *covalent* and *ionic bonds* (see Figure 4.4) [1, 2].

4.1.3 Chemical Potential

Every species strives for the lowest state of energy. For example, if you place a ball on a table, it will fall to the floor due to gravitation when it is pushed over the edge. Thus, the ball energetically prefers to be on the floor.

A molecule will move from a state of higher energy to a state of lower energy in the same manner as the ball. The energy state is denoted chemical potential where a molecule will always move to a state of lower chemical potential. This can take place either through diffusion or by a chemical reaction, that is diffusion reduces concentration differences, which reduces the chemical potential, and molecules react if it is energetically more preferential [2].

4.1.4 Surfaces

The solid surfaces in tribological contacts can be made of any materials, for example metals, ceramics, polymers or elastomers. The performance of the lubricated contact is also affected by the existence of other liquids in the lubricant (e.g. water) or gas in the lubricant (e.g. air). Hence, there are several types of surfaces in the lubricated contact. Besides the solid–solid surface, surfaces of importance are solid–liquid, liquid–liquid, gas–liquid or solid–gas–liquid surfaces (see Figure 4.5) [2, 3].

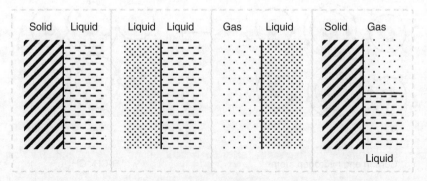

Figure 4.5 Surfaces that may occur in lubricants and in lubricated contacts

A *solid–liquid surface* describes the contact between the lubricant and the solid bodies, including the contact with, for example, soot or wear particles. The contact between a solid phase and a liquid phase of a lubricant, which occurs, for example, during wax crystallization at temperatures close to the pour point, can also be considered as a solid–liquid surface.

A *liquid–liquid surface* describes, for example, water entrainment, where water and lubricant are brought into contact with each other. Water entrainment may enhance oxidation of the lubricant, give rise to hydrolysis and promote corrosion. Thus, it is important to understand the character of this liquid–liquid surface in order to control it.

A *gas–liquid surface* describes, for example, air entrainment and foaming in a lubricant, but also the contact between the lubricant and the free surface of air above it. In all cases air and lubricant are brought into contact with each other. Air entrainment will affect both air release and oxidation. The free surface of air may to a certain extent affect oxidation and may at high temperatures give rise to evaporation of the lubricant. The gas–liquid contact is undesirable and it is important to understand how to minimize it.

A *solid–gas–liquid surface* describes where air (or oxygen), water and a metal surface are brought into contact with each other. The air and water content should preferably be reduced in the lubricant, as has been stated above. Also, the combination of air and water may cause corrosion of the metal surface.

For chemical reactions or physical interactions to take place at any surface, the additive chemistry must allow interaction across the surface. Thus, additive molecules must be transferred from the lubricant bulk to and across the surface, which brings the theory to mass transfer.

4.1.5 Mass Transfer

In the bulk region of the lubricant, turbulent mixing most likely takes place and molecular transport occurs through turbulent eddies. The mixing changes character at the vicinity of a surface. Close to a surface, a thin boundary layer of lubricant is formed that is more stagnant than the bulk. Here, molecular transport occurs only through diffusion. Intense mixing in the bulk may reduce the thickness of the boundary layer, but even with very intense mixing the boundary layer still exists [3–5].

Molecules move from a state of higher chemical potential to a state of lower chemical potential. In practice, this means that concentrations even out with time (see Figure 4.6). For instance, when changing lubricants in a system via top-up of a new lubricant, they will mix by themselves even if not mechanically mixed. If they are mixed via mechanical stirring the mixing time will be shortened.

Fick's first law describes mass transfer via diffusion:

$$J = -D\frac{\partial c}{\partial x} \qquad (4.2)$$

where the mass transfer rate J depends on the diffusivity (or diffusion coefficient) D of the diffusing molecule and the difference in concentration c across a thickness x. Thus, mass transfer increases if the concentration difference is increased, if the thickness of the boundary layer is decreased or if the relative diffusivity is increased.

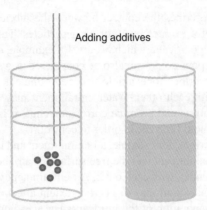

Figure 4.6 Fick's first law of diffusion is visualized for adding additives during, for example, production to a base fluid at time 0 (to the left) and time t (to the right)

4.1.6 Adsorption

Adsorption describes the phenomenon of molecules being attracted to a surface. Such attraction depends both on the chemistries of the adsorbing molecule and the surface, and on the base fluid chemistry. Attraction is promoted if the chemistries of the adsorbing molecule and the base fluid are similar enough to allow for blending, but different enough for the adsorbing molecule to prefer the surface to the base fluid.

When estimating the rate of adsorption of molecules to a surface (see Figure 4.7), the following assumptions are made:

- The surface has a fixed number of adsorption sites. At equilibrium (at any temperature), the fraction θ of the sites is occupied by adsorbed molecules, while the fraction $(1 - \theta)$ is not occupied (i.e. these sites are free).
- Each site can hold only one molecule.
- The adsorption energy is the same for all sites and does not depend on θ.

Molecules may both adsorb and desorb from the surface. The rate of adsorption v_a is

$$v_a = k_a c_{solution}(1 - \theta) \tag{4.3}$$

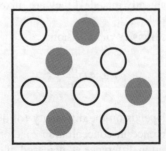

Figure 4.7 Molecule A covers 40% of the active sites, that is, θ_A is 0.40 and $1 - \theta_A$ is 0.60

where k_a is the rate constant for adsorption and $c_{solution}$ is the concentration of the adsorbing additive in the lubricant. The rate of desorption v_d is

$$v_d = k_d c_{surface} \theta \qquad (4.4)$$

where k_d is the rate constant for desorption and $c_{surface}$ is the concentration on the surface. The fraction of occupied sites θ reaches an equilibrium, that is when the rate of adsorption equals the rate of desorption, $v_a = v_d$, which gives the Langmuir adsorption isotherm

$$\theta = \frac{\frac{k_a}{k_d} c_{ads}}{1 + \frac{k_a}{k_d} c_{ads}} = \frac{K c_{ads}}{1 + K c_{ads}} \qquad (4.5)$$

where K (i.e. k_a/k_d) is the coefficient of adsorption and c_{ads} is the ratio of $c_{solution}$ and $c_{surface}$ (i.e. $c_{solution}/c_{surface}$).

The affinity for the surface depends on the combination of the adsorbing molecule and the surface chemistry. Adsorption occurs since it is energetically more beneficial for the surface active additives to adsorb to the surface instead of being in the lubricant bulk. The number of sites on the surface will limit the amount of additives that can be adhered to it, allowing a certain packaging density on the surface.

Surface active molecules are commonly depicted by one polar moiety and one hydrocarbon chain. However, the packing density can be reduced if the hydrocarbon chains of the adsorbing molecule are multiple or branched since they will occupy a larger space. For example, the copper passivator tolutriazole molecule has three hydrocarbon chains pointing away from the polar moiety (see Figure 4.8). These three hydrocarbon chains occupy a certain spacious volume, thereby hindering other surface active additives to adsorb to neighbouring active sites on the surface. This process is called steric hindering.

Adsorption to surfaces can be done either by weak forces (i.e. physical adsorption or physisorption) or stronger forces, giving rise to atomic bonds (i.e. chemical adsorption or chemisorption).

Physisorption involves van der Waals forces and may occur at low temperatures. It may give rise to the formation of multilayers since no chemical bonds form. It is in general reversible and physisorbed layers can be removed by washing the surface with strong solvents.

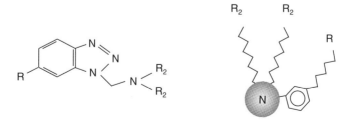

Figure 4.8 The tolutriazole molecule: the general structure (left) and a model cartoon drawing (right)

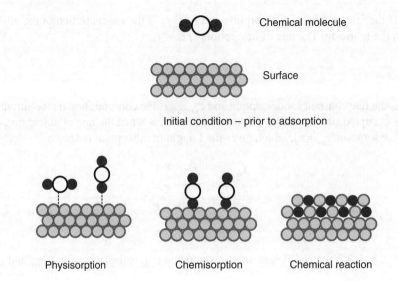

Figure 4.9 Visualization of physisorption, chemisorption and a chemical reaction on a surface. Physisorption involves weak bonding and chemisorption strong bonding. Chemical reaction changes the surface composition (For a colour version of this figure, see the colour plate section)

Chemisorption occurs when the adsorbing molecule interacts more strongly with the surface. After the initial adsorption, bonds within the adsorbing molecule are broken allowing new bonds (either ionic or covalent bonds) to form with the surface. This usually requires energy to take place. It always gives rise to monolayers. Chemisorption is irreversible and chemisorbed layers are not possible to remove, even with strong solvents.

Sometimes, chemisorption is followed by *chemical reactions* resulting in a layer of a material compound on the surface (see Figure 4.9) [2, 5].

4.1.7 Chemical Characteristics of Surface Active Additives

The surface active additives have a polar moiety and an oil-soluble hydrocarbon chain (see Figure 4.10). The polar moiety is the active part of the additive. The hydrocarbon chain length determines the solubility of the additive in the lubricant. A long hydrocarbon chain allows for good solubility with the base fluid, while a short hydrocarbon chain is preferred for the additive to migrate in the base fluid and be surface active in the application.

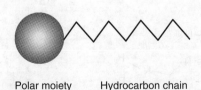

Figure 4.10 A generalized structure of a surface active molecule with a polar moiety and a hydrocarbon chain

Additives

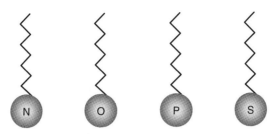

Figure 4.11 The polar moiety of surface active additives typically consists of nitrogen, oxygen, phosphorus or sulfur. These atoms are all prone to attract electrons from the surroundings, giving the molecule its polar entity

Polarity implies that the molecule is asymmetrical and has a different chemical affinity at the two ends of the molecule. The polar part, the polar moiety, commonly consists of nitrogen, oxygen, phosphorus and/or sulfur chemistry (see Figure 4.11). These are all preferred chemistries for many surface active additives. These atoms all lack electrons in their outmost shell, making them prone to attract electrons from neighbouring atoms, that is the polar moiety is negative.

The additive chemistry must allow for dissolving in the base fluid and the surface active additives must possess certain properties to adsorb and react with the surface. For surface active additives to function properly, the additive must be transferred to the vicinity of the surface, the chemical potential must promote the transfer of the additive to the surface, there must be available sites to adsorb to and the net adsorption must be positive. For additives that should react chemically with the surface to function properly, the energy must be high enough to favour reaction and the reaction rate must be high enough to proceed.

In an application the adsorption layers may be removed when exposed to, for instance, high loads and may be formed again when the adsorbing molecules are brought into contact with the surface. The formation rate must be higher than the removal rate in order to protect the surface.

4.2 Additive Exploration

The additives will be described according to their action mechanism in the lubricant (see Figure 4.12). The exploration will cover a general description, the acting mechanism, and the additive chemistry. The additives perform either on a surface or in the lubricant bulk. The surfaces involved are solid–liquid, liquid–liquid and liquid–gas. The additives in the bulk are surrounded by base fluid molecules and are assumed to be able to freely move in all directions. The bulk active additives may either physically interact in the lubricant or they may react with the molecular species in it.

Some concepts will be used in the additive exploration and need an explanation before continuing. These include polymers and molecular weight. Polymers are large molecules or macromolecules composed of repeating structural units. These units are commonly connected by covalent bonds. The molecular weight is the sum of the atomic weights of the contributing atoms. For example, water has the molecular weight of 18 g/mole, while larger molecules such as polymers may have molecular weights up to 100 000 g/mole.

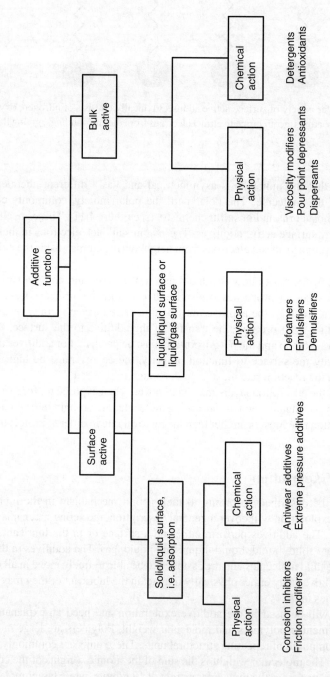

Figure 4.12 Additives are categoriszed and explored according to their type of action in the lubricant

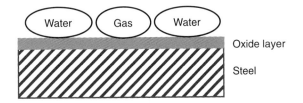

Figure 4.13 A metal surface, here represented by iron, is typically oxidized and covered with water and gas

4.3 Surface Active Adsorbing Additives

Surface active additives acting at the solid–liquid surface are added to the lubricant in order to protect the solid from corrosion and wear and to reduce friction. They are sometimes referred to as film-forming additives. All metals and alloys, with the exception of noble metals, will oxidize in the presence of air. In normal atmosphere, all metal surfaces are covered with water or gas (e.g. air) (see Figure 4.13). The action of surface adsorption involves several steps, such as removal of water and gas, transfer of the additive from the bulk lubricant to the surface and adsorption on the metal surface. Sometimes the adsorbed layer reacts with the surface.

The metal surface, commonly iron-rich steel in tribological contacts, has electrons available for interaction with the surface active additives. In many cases the additives physisorb to the surface, that is there is no chemical reaction. In other cases there is a strong interaction, involving exchange of electrons, resulting in chemisorption of the additives or even in a chemical reaction that forms a layer of a new material compound on the metal surface.

A nascent iron surface is far more reactive than an oxidized iron surface since it has more freely moving electrons available for interaction with additives. The oxide layer is removed during sliding, exposing the nascent surface and allowing numerous sites for reactions. The reaction between sulfur and iron proceeds 1000 times faster for nascent surfaces than oxide-covered surfaces. However, engineering surfaces are in most cases covered with an oxide layer. The lubricant additives will, therefore, interact with the oxidized surface [2].

4.3.1 Corrosion Inhibitors

Corrosion is a disintegration process between a metal surface and the surrounding. The surrounding nature (i.e. the oxide film, presence of acids, water, oxygen and temperature) and the metal itself (i.e. composition, structure and surface roughness) affect the corrosion process. All metals may corrode, apart from the noblest ones, like gold (see Figure 4.14) [6].

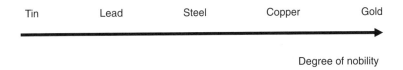

Figure 4.14 The order of nobility of some metals. Metals with lower nobility are more prone to corrode

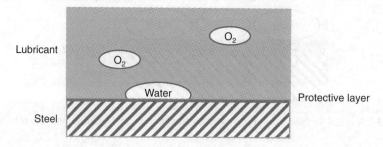

Figure 4.15 Steel corrosion is reduced by the protective layer formed by corrosion inhibitors

Corrosion processes relevant in lubricated contacts involve electrochemical corrosion and chemical corrosion. Electrochemical corrosion occurs in the presence of electrolytes[3] and chemical corrosion occurs without any electrolyte. Some chemical substances such as acids and sulfur will cause chemical corrosion. Corrosion can be both selective and nonselective in alloys. Selective corrosion implies removing one or several metals in the alloy. This may be the case where lead is removed and not copper in alloys containing both metals. Nonselective corrosion processes will remove all metals in the alloy to the same extent.

There are two types of corrosion inhibitors that can be added to the lubricants: acid neutralizers and film-forming additives. The most common ones are film-forming additives. They are surface active and adsorb on the metal surface via mainly physisorption, even though chemisorption may occur. The corrosion inhibitors are chosen because of their high surface affinity for the selected metal.

Steel corrosion occurs at the interface between steel, water and lubricant, in the presence of oxygen. It is electrochemical in its character. For instance, it may take place in combustion engines, especially where the driver often starts and turns off the engine. Water is formed from the combustion, oxygen is introduced from the air and the electrolyte is formed via the reaction of metals with acids that form during combustion or oxidation [6].

The corrosion inhibitor adsorbs to the steel surface, thereby preventing water and oxygen from reaching the surface (see Figure 4.15). They are highly surface active and even if they are removed, the chemical potential strongly promotes fast adsorption of new corrosion inhibition molecules. The chemistry of corrosion inhibitors involves amines (containing nitrogen), neutral and basic sulfonates (containing sulfur), succinic acids (containing a carboxylic acid group, COOH) and phosphites (containing phosphorus). Long chain amines will physisorb on the surface, while the others mentioned will chemisorb.

Yellow metal corrosion (shown by observing copper corrosion) is a chemical type of corrosion. It occurs due to the attack of chemical species on the copper surface. These species may be formed during lubricant oxidation or combustion in an engine.

The yellow metal corrosion inhibitors adsorb by physisorption to the surface, preventing water and oxygen from reaching it (see Figure 4.16). They are extremely surface active and tiny amounts are enough to effectively protect the yellow metal surface from corrosion. Copper corrosion inhibitor chemistry includes film-forming additives such as triazoles (see Figure 4.8) [6].

---

[3] An electrolyte contains ions and can conduct electricity.

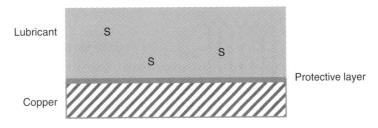

Figure 4.16 Yellow metals are protected by a layer of yellow metal passivators

Soft metal corrosion involves the selective removal of soft metal from alloys (e.g. lead and tin). It takes place due to the aggressive attack of species on the metal surface. No effective inhibitors are known and soft metal corrosion is believed to be minimized by proper balancing of the additives in the lubricant [6, 7].

4.3.2 Friction Modifiers

Friction modifiers (FMs) are added to the lubricant to modify the friction in the mixed lubrication regime of the tribological contact. Adding 1% is enough to give a significant change in friction. They are active in the mixed lubricating regime at moderate temperatures and loads, but will desorb at high temperatures and high loads. To a certain degree this desorption can be compensated for by adding more FMs to the lubricant. They are added, for example, to engine oils and gear oils to reduce friction and to automatic transmission fluids (ATFs) to control friction [6, 8].

Friction modifiers adsorb by physisorption, even though chemisorption occurs. They adsorb in monolayers or multilayers. The FM molecules adsorb more strongly to the surface than to neighbouring FM molecules (see Figure 4.17). The layers will promote the separation of the surfaces by steric hindering, exhibiting strong anticompressive behaviour of the layers. On the other hand, the FM layers are easily sheared and thereby reduce friction. This combination of strong anticompressive properties and low resistance to shear gives FMs their superior properties.

The chemistry comprises, for example, fatty alcohols, esters, fatty acids and fatty amides. These chemistries are all surface active, having a polar part of oxygen or nitrogen and a hydrophobic part.

Long and linear chain materials reduce friction more effectively than short and branched chain FMs. Fatty acids reduce friction more effectively than both fatty amides and fatty alcohols. The preferred fatty acid FM is saturated and has 13 to 18 carbons. Shorter chained fatty acids may entail corrosion problems [8].

4.3.3 Antiwear Additives

Antiwear additives (AWs) prolong the life of the tribological contact by modifying the metal surface and thereby reducing wear in the mixed lubrication regime. They are active in the

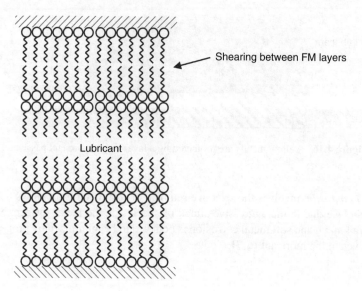

Figure 4.17 Multilayer adsorption of FM additives promotes separation of surfaces. The layers are easily sheared, giving low friction

mixed lubrication regime at higher temperatures and loads than FMs. They are commonly added at 1–3%. They are used in, for example, engine oils, hydraulic fluids and ATFs [8–10].

Antiwear additives mainly chemisorb to metal surfaces. The chemisorbed AW monolayer offers durable wear protection of the surface in the mixed lubrication regime. AWs require moderate activation temperatures, loads and shear rates, but higher than FMs to become activated [9].

The chemistry involves nitrogen, phosphorus and/or sulfur, where phosphorus is the most widely used. Phosphorus offers antiwear protection at relatively low loads. The most frequently used AW is ZDDP (i.e. zinc dialkyldithiophosphate), which also has antioxidant and detergent properties.

Nontraditional AW/EP (extreme pressure) additives will find more extensive usage as environmental awareness increases. Today, there are alternatives reported in the literature. Examples are graphite, PTFE, inorganic carbon nanoparticles and CHON-based additives[4] [9, 10].

4.3.4 Extreme Pressure Additives

Extreme pressure additives modify the metal surface in order to avoid scuffing and control wear in the boundary lubrication regime. They form protective, low shear strength surface films that reduce friction and wear. However, too high concentrations may cause excessive corrosion and wear. EPs are needed in slow moving, heavily loaded gears (i.e. gear oils) and metalworking fluids [7, 9].

---

[4] CHON is an abbreviation of C (carbon), H (hydrogen), O (oxygen) and N (nitrogen).

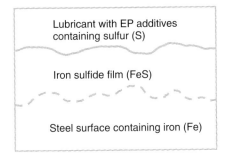

Figure 4.18 The EP additives react with the surface forming an iron sulfide compound layer

There are mild EP and strong EP additives. Mild EPs require lower temperatures than strong EPs to become activated. EP additives require higher activation temperatures, loads and shear rates than AWs.

EP additives initially chemisorb to surfaces. However, high loads, high temperatures and commonly high shear in the boundary lubrication regime will remove the hydrocarbon chains, thereby breaking bonds in the additive structure. The removal of the hydrocarbon chains will allow the reaction between the polar moiety and the metal surface, for example sulfur polar moiety and iron surface forming iron sulfide (see Figure 4.18). The reaction rate is usually high and the film formed has low friction and high scuffing resistance [9].

Different chemistries are used for obtaining EP performance. They are all surface active with a polar moiety of nitrogen, phosphorus, sulfur and/or halogens, where phosphorus and primarily sulfur are predominant. Halogens are less widely used due to their poor environmental properties. Sulfur containing EP additives decompose at high temperatures forming iron sulfide layers. The layer thickness depends on the sulfur content. Phosphorus EP additives react with the metal surface, forming a metal phosphate. The phosphorus additives require higher temperatures than sulfur additives to form a protective layer.

4.3.5 Activation of Antiwear and Extreme Pressure Additives

The action of AW and EP additives is promoted by increased temperatures, pressures and shear. Both adsorb at moderate temperatures and pressures. At high temperatures or pressures they form reaction layers by reacting chemically with the surface. This is visualized in Figure 4.19 showing a body sliding across a flat surface. The area of contact will move (shown as near-contact area, contact area and off-contact area) when the body slides across the flat surface. In the contact area, both temperature and pressure are high, which will lead to the formation of a reaction layer on both surfaces. However, the sliding motion may also result in wear, that is partial removal of the reaction layer.

As the body slides across the surface the contact area will turn into an off-contact area, where the pressure is lower with a possibly remaining high temperature. Adsorption will take place and chemical reaction will proceed if the temperature is high enough. In front of the contact area (in the sliding direction) is the near-contact area, becoming the next contact-area. Friction heat from the contact area may spread due to conduction to the near-contact area. The

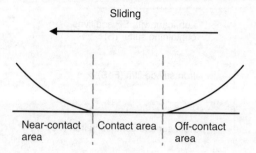

Figure 4.19 Different processes occur depending on the contacting conditions in each specific area during sliding of a circular body across a plane surface

pressure is low and adsorption to the surfaces occurs. The temperature reached via conduction may be high enough to allow reactions with the surfaces to occur [11].

4.3.6 Competition for Surface Sites by Surface Active Additives

Several types of surface active additives may be added to a lubricant. The adsorption of surface active additives depends on the activity of the surface, the temperature, the load and the activity of the additive. The coverage of additives may therefore differ if the lubricant contains only additive A, only additive B or a mix of the two (see Figure 4.20).

This theory can be supplemented by information about FMs, AWs and EPs. The action of these surface active additives is shown in Figure 4.21. In the mixed lubrication regime both FMs and AWs are active, which can give rise to a competition for available surface sites. However, at increased loads FMs are removed, promoting adsorption of primarily AWs.

EP additives require even more severe conditions to be activated, that is high temperatures and high loads, to react with the iron surface. This is the situation in the boundary lubrication regime. Here the competition of EPs with the FMs and AWs is of minor importance.

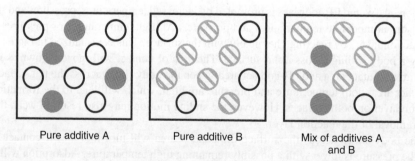

Figure 4.20 Additive A covers 40% of the active sites, that is θ_A is 0.40. Additive B covers 60% of the active sites, that is θ_B is 0.60. In a blend (to the right), additive A may cover 30% of the active sites and additive B may cover 50% of the active sites, leaving 20% empty

Tribology is the science of friction, wear and lubrication. Tribological contacts consist of solid bodies in contact under relative motion with or without any lubrication. Different contact geometries and conditions give rise to different types of contact areas, namely distributed, line and point contact areas. ***The character of a lubricated contact*** depends on the surface conditions, the surface roughness and on how well the lubricant film may separate the surfaces. The surfaces are either fully separated (i.e. full-film lubrication), in continuous contact (i.e. boundary lubrication) or occasionally in contact (i.e. mixed lubrication).

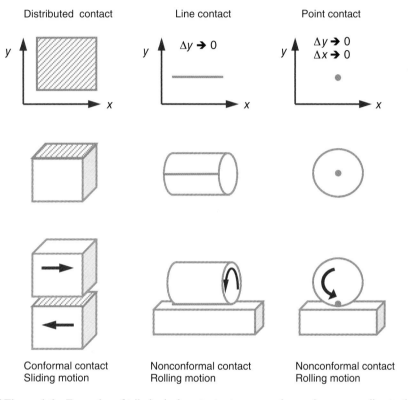

Plate 1 / Figure 1.6 Examples of tribological contacts at macroscale are shown according to the type of contact area

---

Lubricants: Introduction to Properties and Performance, First Edition.
Marika Torbacke, Åsa Kassman Rudolphi and Elisabet Kassfeldt.
© 2014 John Wiley & Sons, Ltd. Published 2014 by John Wiley & Sons, Ltd.

The lubricant properties should meet various requirements. The performance properties are crucial for the action of the tribological contact. The life of a lubricant is affected by the choice of base fluids, and also on the action of certain added additives. A product with good environmental properties can be designed by a clever selection of base fluids and additives. Such a product will reduce the impact both on short and long terms for humans and the surrounding environment.

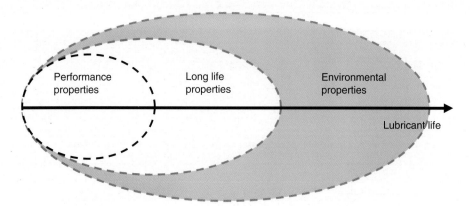

Plate 2 / Figure 2.1   The timescale of different lubricant properties

The lubricant properties are measured and verified during development, in production and in real applications. ***Viscosity*** is the key parameter for lubrication of a tribological contact, serving as the basis for the lubricant film thickness.

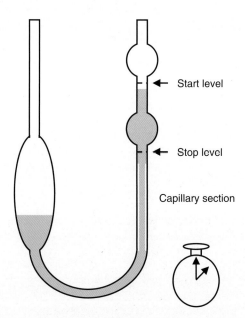

Plate 3 / Figure 2.4   Kinematic viscosity is measured in a capillary viscometer. The fluid flows due to gravitation and the time for a certain amount of fluid to flow through the capillary section is measured

Base fluids originate from different sources. They can be from natural gas, coal, crude oil or from plants and animals. The source will govern the final properties of the base fluids. Base oils from crude oil show good lubricating properties and by treating them with hydrogen the life can be prolonged. Base fluids from plant and animal sources show good lubricating properties and in addition they have inherent biodegradable and renewable properties, making them suitable for environmentally adapted lubricants. The base fluid itself will provide the lubricated contact with most of the properties needed for good lubrication.

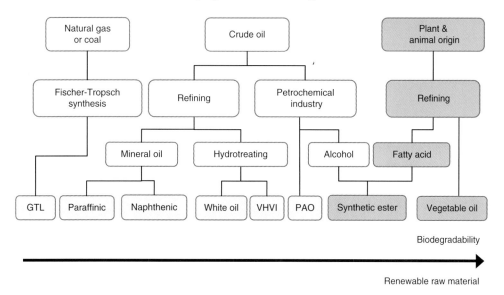

Plate 4 / Figure 3.3 The origin of different base fluids

Synthetic esters may symbolize environmentally adapted lubricants. Synthetic esters allow development of technically advanced lubricants with reduced impact on the environment both in the short term and long term. They are inherently biodegradable and renewable to a degree dependent on the used raw materials.

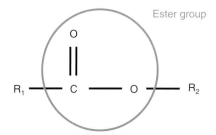

Plate 5 / Figure 3.9 A schematic of a monoester, where COO is the ester group

Additives are added to the base fluids to enhance the properties of the lubricated contact. This may comprise reduced friction and wear during boundary and mixed lubrication as well as prolong lubricant life by enhanced antioxidancy properties. Additives may be surface active and form surface films by interacting with the solid body surfaces. The interaction depends on the contact conditions. Sometimes a ***competition for active surface sites*** arises, which needs to be understood during formulation.

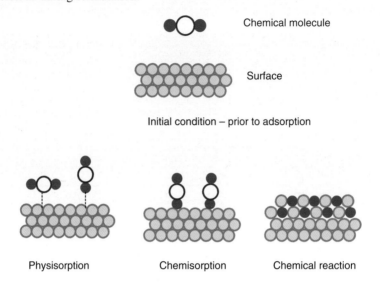

Plate 6 / Figure 4.9 Visualization of physisorption, chemisorption and a chemical reaction on a surface. Physisorption involves weak bonding and chemisorption strong bonding. Chemical reaction changes the surface composition

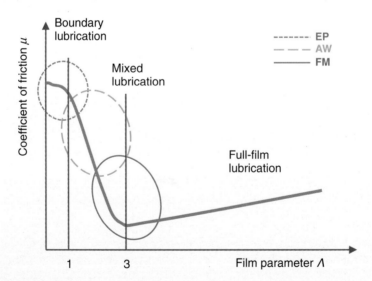

Plate 7 / Figure 4.21 The lubrication regimes where friction modifiers (FMs), antiwear (AW) additives and extreme pressure (EP) additives are active

Each lubricant is a refined mix of base fluids and additives, providing the final lubricant with its properties. The formulation depends on the application requirements, and also on marketing intention. Mainline products are commonly based on paraffinic base oils, premium products on synthetic base oils and environmentally adapted lubricants on natural or synthetic esters. *Lubricant properties can affect friction in the contact* by, for example, lowering it throughout the working area or in the boundary lubrication only. This should be considered and tested during development.

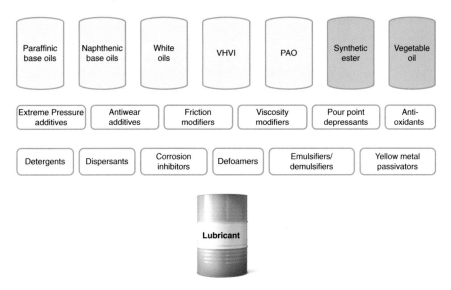

Plate 8 / Figure 5.1 Lubricants can be formulated with a variety of base oils and additives

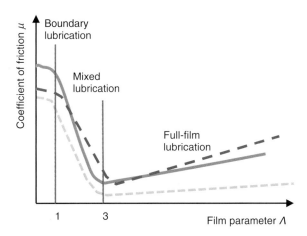

Plate 9 / Figure 5.4 Two examples of lubricants showing better performance than the reference lubricant

Testing supports development by verifying properties. Important functionality to test includes film thickness, friction and wear. The film thickness serves as the basis for the separation of the surfaces in the lubricated contact. Friction and wear govern the performance and life of the lubricated contact. The selection of test methods is based on the geometry of the real application contact as well as the anticipated lubrication regime for operation.

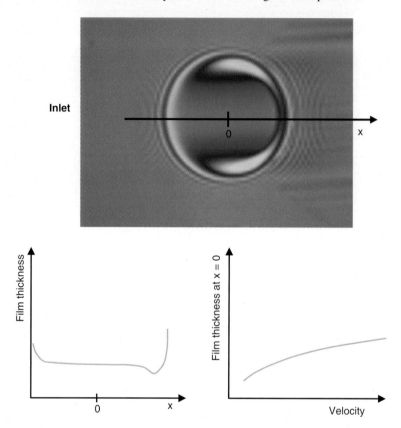

Plate 10 / Figure 6.6 Interferograph (interferometry photograph), calculated lubricant film thickness from the interferograph (bottom left) and film thickness as a function of velocity (bottom right) calculated from several interferographs

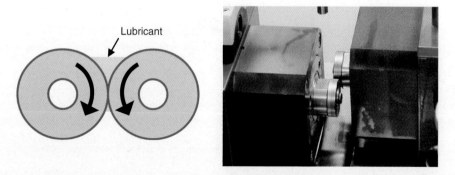

Plate 11 / Figure 6.13 Illustration of the twin disc set-up. Close up view of the test specimen in the twin disc set-up. Lubricant may be dropped on to the discs

Lubricant characterization ensures product quality and functionality in the application. Basic analyses will prove the most important characteristics of the lubricant. It can be used to investigate failures and/or indicate the relevant time for lubricant change. It is crucial to collect a representative sample prior to analysing. A representative sample will form the basis for a correct and relevant analysis result.

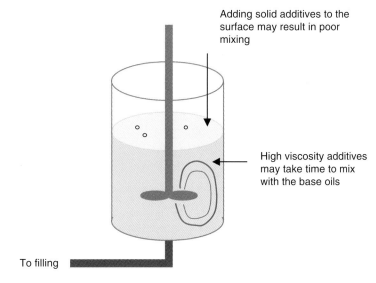

Plate 12 / Figure 7.2 Visualization of areas where sampling may be difficult

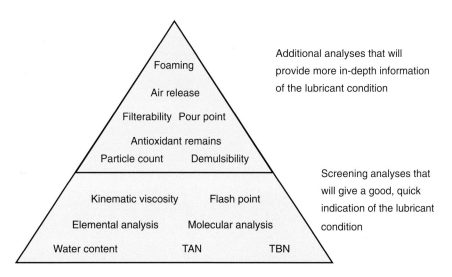

Plate 13 / Figure 7.5 Examples of lubricant analyses

Surface characterization supports the release of new mechanical components, the understanding of test results and issues arising in real applications. Surface characterization comprises microscopy, surface measurements giving roughness information, hardness measurements and surface analysis. It can be performed on surfaces and cross-sectional areas, providing information on friction and wear mechanisms as well as on physisorbed, chemisorbed or reacted layers.

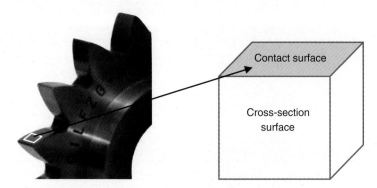

Plate 14 / Figure 8.2 Schematic of a piece of material cut out from a component. Both the original contact surface and the cross-section surfaces are of interest to characterize

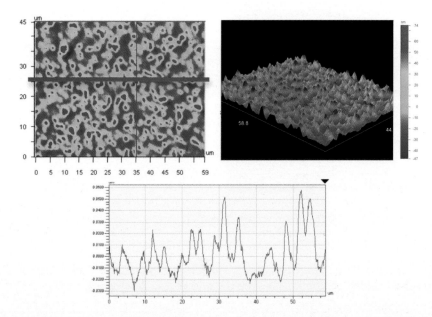

Plate 15 / Figure 8.3 Examples of VSI results. The size of the imaged area is 59 µm times 45 µm. Colours (or a grey scale) illustrate the height of the surface in the 2D (upper left) and 3D (upper right) images, the scale ranging from –47 to 74 nm. The position of the line profile (bottom) is indicated in the 2D image

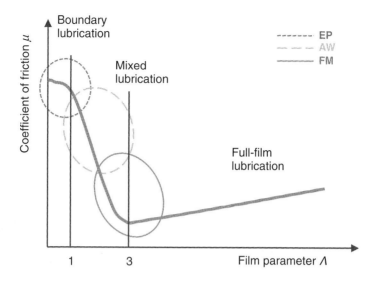

Figure 4.21 The lubrication regimes where friction modifiers (FM), antiwear (AW) additives and extreme pressure (EP) additives are active (For a colour version of this figure, see the colour plate section)

In all lubricating regimes also the corrosion inhibitors are surface active and they are very selective for the metals they are designed for (e.g. steel or yellow metals). Having corrosion inhibitors in excessive amounts may therefore reduce the effect of FM, AW and EP additives in the lubricated contact.

The competition for the metal surface is of course considered during lubricant formulation. It is important not to use excessive amounts during formulation, since this may destroy the action of other surface active additives, besides bringing up the price of the final lubricant.

4.4 Interfacial Surface Active Additives

Surface active additives may also act on liquid–liquid or liquid–gas surfaces. This is the case for defoamers, emulsifiers and demulsifiers. Defoamers act on liquid–gas surfaces, while emulsifiers and demulsifiers act on liquid–liquid surfaces.

4.4.1 Defoamers

Defoamers have low solubility in lubricants and are very surface active. They are added to almost all lubricants to avoid foam formation, foam growth or cavitation. Thus, they are added to increase lubricant availability as well as controlling foam and air entrainment [7].

Foam consists of air encapsulated in a thin film of lubricant (see Figure 4.22). This film is elastic and the air pressure inside the bubble is in balance with the pressure outside the bubble. Gravity acts on the lubricant film, but this is counteracted in stable foams and during foam build-up. The surface active defoamers will act at the lubricant–air surface, with the polar moiety pointing to the air and the hydrocarbon chain remaining in the lubricant. In this

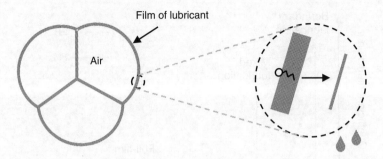

Figure 4.22 Foam visualized by three bubbles to the left and the action of a defoamer causing film thinning to the right

way, defoamers destabilize the foam by decreasing the film thickness. At a certain thickness, this will cause film rupture, which is further enabled by gravitation draining the film. Finally, bubbles coalesce and the foam collapses, resulting in droplets of lubricant instead of bubbles of air. Defoamers will counteract the action of some foam stabilizing additives such as detergents, dispersants and viscosity improvers.

The chemistry of the defoamers guarantees liquid/air contact, that is one part of the additive molecule prefers air and the other liquid. They are quickly dispersed and distributed in the lubricant. The most frequently used defoamers are based on silicone oils or polymethacrylates. They are added in very small amounts, that is parts of a percentage.

4.4.2 Emulsifiers and Demulsifiers

Emulsions are desired and commonly used in metal-working applications where water is used for cooling purposes. Since the heat capacity of water is significantly higher than the heat capacity of oil, water will remove heat from the contact more quickly than oil. In other applications emulsions are not desired, but may unintentionally form in the lubricant when water is entrained. Both emulsifiers and demulsifiers are used to control the amount of water carried by the oil [7].

Emulsifiers are used when entrained water cannot be separated from the lubricant within the system. This may be due to high circulation rates, which does not allow time for water separation. They may be added to engine oils and metal-working products (e.g. rolling fluids). Emulsifiers reduce the surface tension of water. Therefore, they facilitate the formation of an emulsion, that is they divide the water volume into smaller droplets. They commonly have molecular weights of 2000 g/mole or less and they have a hydrophilic group based on nitrogen, oxygen, phosphorus or sulfur.

Demulsifiers act in the opposite way to emulsifiers and improve water separation. They increase the water surface tension, which will increase the water droplet size until the water droplets sink. The removal of water may be important to reduce hydrolysis or oxidation of the lubricant. They are used when rapid separation of water is needed. This is the case for ATFs, hydraulic oils and industrial gear oils. Demulsifiers are polymers with molecular weights of up to 100 000 g/mole and they contain 5–50% polyethylene oxide. They have a hydrophilic group based on nitrogen, oxygen, phosphorus or sulfur.

4.5 Physically Bulk Active Additives

Bulk active additives are usually large macro molecules, which tumble more slowly in the lubricant than small molecules (i.e. steric action). They interact physically with other compounds in the lubricant.

4.5.1 Viscosity Modifiers

Viscosity modifiers (VMs) or viscosity index improvers are bulk active additives that alter the viscosity of the base fluid through steric action. They are added to lubricants performing over a broad temperature range. Single-grade engine oils contain no viscosity modifiers and must therefore be changed twice a year (i.e. every spring and autumn). Today's multigrade engine oils contain VMs and do not have to undergo seasonal oil changes [7, 12, 13].

The ideal VM should reduce the fuel consumption (i.e. improve fuel economy), be added in low concentrations, not interact with other chemistries (including both base fluids and additives) and be shear stable.

VMs increase the lubricant viscosity at all temperatures. However, the viscosity increase is most significant at higher temperatures where the polymer chains extend due to the added thermal energy. At lower temperatures the viscosity modifier occupies only a small volume (see Figure 4.23). Thus, the solubility of the polymeric chain increases with increasing temperature and the VM thereby associates more with the base fluid molecule. Consequently, the viscosity index of the base fluid increases.

They are polymers with average molecular weights of 10 000–150 000 g/mole. They are the largest molecules in the lubricant. The thickening efficiency is proportional to the molecular weight or the length of the polymer chain. The longer the polymer chain, the greater is the thickening efficiency. The lower weights are more common since they show superior shear stability properties. The selection of VMs depends on the application, the base fluid characteristics, the viscosity grade needed, the shear stability requirement, the

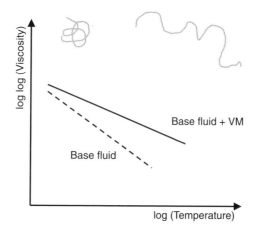

Figure 4.23 Viscosity modifiers (VMs) increase the viscosity at high temperatures and consequently increase the viscosity index and thereby reduce the lubricant viscosity dependence on temperature

Figure 4.24 A viscosity modifier (VM) drawn as a cartoon structure. Each line in the drawing represents a hydrocarbon chain

low temperature viscosity requirements, the ability to be dispersed in the lubricant and the economics. The stability is important since they may lose in viscosity due to shear or due to ageing (i.e. polymer degradation).

The chemical structure can be described as a polymer chain with short hydrocarbon branches (or side chains) (see Figure 4.24). They are manufactured from either *olefin-based polymers* (e.g. polyisobutylene (PIB), olefin copolymers (OCPs)) or *ester polymers* (e.g. polyalkylmethacrylate (PMA)).

PIBs were the first known VMs and were very important in the past for engine oils. PIBs have a comparatively simple molecular structure. Their low temperature properties are poor, but they are inherently shear stable.

OCPs are polymers with a fairly simple chemical structure. They are the most important VMs and are commonly used in engine oils. They show good thermal stability and perform well over a broad temperature range. They are cost effective and have a high thickening efficiency. However, they must be used in combination with a pour point depressant [12].

PMAs are ester-based with a rather complex chemical structure showing good compatibility with many base fluids. They have good low temperature properties, but show only a moderate viscosity-improving ability, due to a low thickening efficiency.

PMAs are superior to OCPs when it comes to oxidative and thermal stability as well as low temperature properties. However, OCPs are considered to be superior if price performance is taken into account. Therefore, PMAs are used when they are technically needed, since they are expensive and used at high treat rates (i.e. high concentrations) to obtain the desired viscosity modification [13].

4.5.2 Pour Point Depressants

Pour point depressants (PPDs) were initially introduced to modify and control wax crystallization phenomena in paraffinic mineral oils. Today, there are also PPDs to lower pour points in other base fluids. However, different base fluids require different PPDs to improve cold flow properties. A good PPD may lower the pour point by as much as 40 °C. There is almost always an optimum concentration level. Above and below this concentration the PPD becomes less effective. They are used in engine oils, ATFs, automotive gear oils and industrial hydraulic fluids [7, 14].

Lubricant viscosity (i.e. the liquid phase) increases when cooling the lubricant. At a certain point the lubricant becomes hazy or cloudy (i.e. the cloud point). This occurs when crystallization of the liquid commences, resulting in a rapid viscosity increase. PPDs will delay this

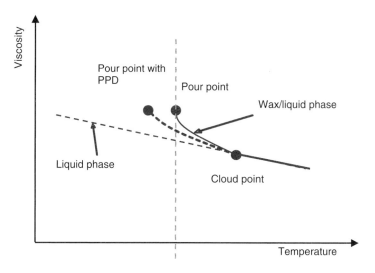

Figure 4.25 The pour point depressant (PPD) will reduce the pour point of the lubricant

rapid increase in viscosity (see Figure 4.25). They are bulk active and lower the pour point through steric action (see Figure 4.26). There are several theories on how they work. They may either alter the crystallization process by inhibiting growth or may increase the solubility of the wax crystals. For example, alkylaromatic PPDs adsorb on to the wax crystals, inhibit crystal growth and improve the pour point in this way [7, 14].

They have a polymeric comb structure with a wide molecular weight distribution (see Figure 4.27). They are commonly made from alkylated wax naphthalenes or polymethacrylates, just like VMs. However, the PPD polymer chain is shorter, while the branches (or side chains) are longer. Thus, their chemistries are similar, but the spatial distribution varies.

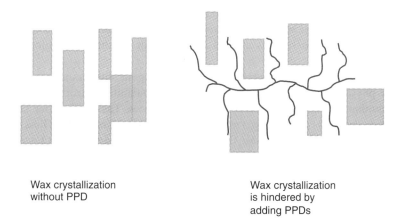

Figure 4.26 The pour point depressant (PPD) hinders crystallization of the lubricant (shown as rectangles) through steric action

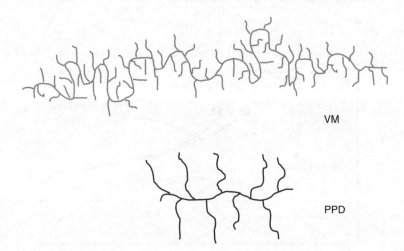

Figure 4.27 The spatial extension differs between viscosity modifiers (VMs) and pour point depressants (PPDs)

Tailoring a PPD involves altering the length, the number and the distribution of side chains, and the nature of the polymeric chain. In general, polymeric PPDs act better than monomeric ones even though high molecular weights are not necessary.

4.5.3 Dispersants

Dispersants prolong the life of the lubricant by dispersing sludge, suspending soot, reducing deposit formation and keeping parts clean. They are added to lubricants used in different applications. For instance, they maintain cleanliness and to a certain extent act as antioxidants in ATFs. They suspend dirt, solubilize AW additives and prevent corrosion in gear oils. In engine oils, they keep residues from the combustion process dispersed in the lubricant, thereby reducing viscosity increase, and improve filterability. Thus, dispersants are used in different applications for cleanliness reasons as well as preventing contaminant agglomeration and sludge [15].

Dispersants are bulk active and perform by dissolving polar contaminants (or dirt) in the lubricant by physical action. Dispersants prevent the contaminants from adhering to each other (i.e. forming aggregates) by adsorbing to their surfaces. The hydrocarbon chains point outwards when adsorbed (i.e. forming a micelle structure) (see Figure 4.28). Thus, dispersants are identical to surface active molecules, that is they have a polar part and a hydrocarbon chain, even though they act in the bulk.

The hydrocarbon chains are long molecular structures requiring a relatively large amount of space around the contaminant particle. These hydrocarbon chains will disturb the attractions between contaminant particles and prevent agglomeration because of the hydrocarbon steric effects via steric repulsion [6, 14].

Typical dispersant chemistry is based on high molecular polyisobutene succinimide molecules with relatively high molecular weights of 1000–2000 g/mole, allowing them to

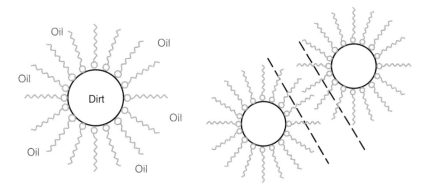

Figure 4.28 Dispersants capture solid contaminants in a micelle structure to the left. Steric repulsion between two micelles of dispersants to the right

be effectively dispersed in the bulk. They have a long hydrocarbon chain and a weak polar moiety containing nitrogen, oxygen, phosphorus or sulfur [15].

4.6 Chemically Bulk Active Additives

Some bulk active additives, such as detergents and antioxidants, react chemically with species in the lubricant. Detergents react with contaminants or combustion residues in order to keep engines clean, thus having a similar function to dispersants in the lubricant. Detergents occur in two different types: neutral and overbased. The neutral detergents are surface active, while overbased detergents are bulk active. However, all detergents are covered together due to their chemical similarities. Antioxidants react with molecules that are formed because of the ageing mechanism of the lubricant, consequently prolonging the life.

4.6.1 Detergents

Detergents are used in high temperature applications where sludge and deposits form due to oxidation or combustion. They are, for example, used in engine oils where they prolong the life of the lubricant by keeping internal engine parts clean. They neutralize the acids formed during combustion, reduce deposits, and thereby prevent sticking of piston rings. Detergents also inhibit oxidation, even though it is a secondary function.

The TBN number of a used lubricant indicates the amount of the remaining detergent concentration. As mentioned in Chapter 2, the TBN number is monitored for engine oils, and a reduction in the TBN is an indicator of when to change engine oils [7, 16].

The two different detergent types are both oil-soluble metal salts produced by reacting a metal base with an organic acid. This can be written as:

Alkaline metal ion + organic acid → oil soluble alkaline metal salt + water

If the alkaline metal ion and the organic acid are added in stoichiometric proportions, the oil soluble alkaline metal salt (i.e. the detergent) is considered to be neutral. If, on the other hand, alkaline metal ions are added in excess (in relation to the amount of organic acid) the detergent is denoted overbased. Both types of detergents will be presented below.

Neutral detergents are primarily used to keep surfaces clean. They are surface active with a relatively short hydrocarbon chain, making them prone to act on metal surfaces in, for example, combustion engines. Detergents are more surface active than most polar components formed in the engine oil and will reduce the adsorption of polar contaminants and thereby corrosion. They may compete for the surface with, for example, AW additives, which must be considered during lubricant formulation. The neutral detergent should be active at high application temperatures.

Overbased detergents are intensively used in, for example, engine oils. The core of the oil-soluble alkaline metal salt will react with acids formed during combustion, thus neutralizing the lubricant. This is possible since the overbased detergents are dissolved in the bulk, allowing a reaction between acids and the detergent core. The overall structure is visualized in Figure 4.29. Lubricants may consist of as much as 50% metal carbonate detergents and still be completely bright and clear. However, typical values are 10% detergents in fully formulated engine oils. In addition, these detergents have antioxidancy properties.

Detergents have short hydrocarbon chains with a strong polar moiety with molecular weights of 150–300 g/mole. The most commonly used alkaline metal ions are calcium and magnesium ions. The detergent chemistry primarily involves sulfonates, phenates and salicylates. The most common detergents are *calcium sulfonates*, which have excellent detergency properties. They can be overbased to yield high base contents. One of their primary advantages is their cost effectiveness. However, they show no antioxidancy properties. *Phenates* and *salicylates* are sulfur free, have good detergency properties and act as antioxidants. *Salicylates* have stronger antioxidative properties than phenates. Unfortunately, neither phenates nor salicylates can reach the same base levels as sulfonates and both are less cost effective [16].

Dispersants and detergents are both surface active molecules with a polar moiety and a hydrocarbon chain. Nevertheless, there are differences between the two. The dispersant polar moiety is weaker than the detergent polar moiety and the hydrocarbon chain is longer

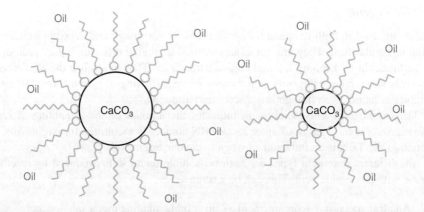

Figure 4.29 An overbased detergent with a core of $CaCO_3$ is shown. The core of the overbased detergent neutralizes acidic components in, for example, engine oils, and is continuously consumed, visualized to the right

Additives

for dispersants than detergents. Consequently, dispersants are more easily dissolved in the lubricant, whereas detergents are prone to be more surface active. Dispersants are ashless (i.e. without metals), while detergents are formulated with metal ions.

4.6.2 Antioxidants

Antioxidants are added to the lubricant in order to slow down the rate of oxidation. A good combination of antioxidants will effectively prolong the life. However, it is crucial for lubricant longevity to reduce the contact with air, lower the temperature and minimize the amount of pro-oxidant wear metal ions in the lubricant. Increased longevity reduces the need for frequent oil changes [7, 17].

The oxidation process is very complex (see also Figure 2.17). For simplicity, it can be said to involve three steps: initiation, propagation and termination. *Initiation* starts the oxidation process when oxygen collides with the hydrocarbon chain (RH), forming radicals (marked with a dot after the chemical short name) and carboxylic acids (ROOH). Radicals are highly reactive chemical species. Carboxylic acids attack metal surfaces, generating carboxylic salts and increasing the rate of oxidation. The *propagation* step involves radical attack on hydrocarbons, forming new radicals. The radical reaction is a chain reaction, which gives rise to an exponential amount of radicals. The oxidation process can be *terminated* either via a radical–radical reaction or through the action of antioxidants.

The use of antioxidants will reduce lubricant oxidation, reduce varnish formation caused by insoluble components and reduce copper and soft metal corrosion initiated by, for example, acids. There are three types of antioxidants: metal deactivators, radical scavengers and hydroperoxide decomposers.

Metal deactivation is done by salicylic acid derivates. They are added in very low quantities (0.5–1%).

Radical scavengers (i.e. primary antioxidants) convert radicals to alcohols and ring-stabilized radicals (see Figure 4.30). The chemistries used are aromatic amines or hindered

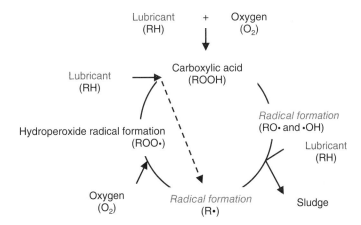

Figure 4.30 Radical scavengers react with radicals formed (marked in italic)

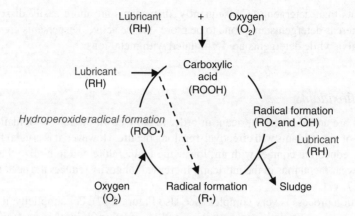

Figure 4.31 Hydroperoxide decomposers react with hydroperoxide radicals formed (marked in italic)

phenols. Aminic radical scavengers are cheaper and more reactive. Phenolic radical scavengers are added to reach the required performance levels. They have low volatility, a long lifetime and are used in small quantities (0.5–1%).

Hydroperoxide decomposers convert hydroperoxides to alcohols (see Figure 4.31). This is done by sulfur compounds, phosphorous compounds (e.g. ZDDP, or zinc dialkyl dithiophosphate[5]) or phenols [7, 17].

4.7 Additive Summary

This chapter has covered additives used in lubricants. Some are surface active and some are bulk active. However, some that are bulk active also show surface active properties. All additives are synthesized to be soluble in the lubricant, although some are designed to be low soluble in order to make them prone to seek out a surface, for example lubricant–air.

Many additives act on metal surfaces. They all have a polar moiety that is prone to interact with the surface. These additives will compete for the active sites on that surface. This is considered during formulation, where the aim is to find the optimum combination providing a balanced formulation fulfilling the application requirements.

Additives covered in this chapter have been characterized according to their solubility and their polarity (see Figure 4.32). They are sorted into two encircled groups: bulk active (i.e. easily dissolved) and the highly surface active additives.

Additives may be purchased as liquids or solids. They may be pure substances or premixed with a small amount of base fluid. Several additives may be blended together (referred to as a package) before being blended with base fluids forming the fully formulated lubricant. Consequently, it is more difficult to give a listed overview of additives and their corresponding properties.

---

[5] ZDDP has antioxidant properties, but is also used extensively as an antiwear additive.

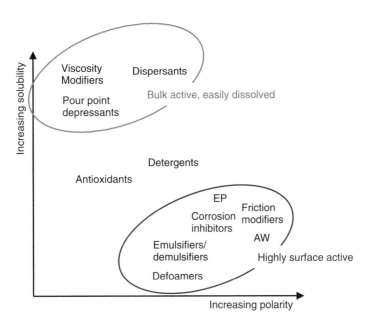

Figure 4.32 Additives have been plotted according to their inherent solubility in lubricants and their polar properties

References

[1] Solomons, T.W.G. (1996) *Organic Chemistry*, 6th edn, John Wiley & Sons.
[2] Moore, W.J. (1983) *Basic Physical Chemistry*, Prentice-Hall International Edition.
[3] Bird, R.B., Stewart, W.E. and Lightfoot, E.N. (1960) *Transport Phenomena*, Wiley International Edition, John Wiley & Sons.
[4] Perry, R.H. and Green, D. (1984) *Perry's Chemical Engineers' Handbook*, 6th edn, McGraw Hill International Edition.
[5] Coulson, J.M. and Richardson, J.F. (1996) *Chemical Engineering Volume 1: Fluid Flow, Heat Transfer and Mass Transfer*, 5th edn, Butterworth-Heinemann, Oxford.
[6] Rizvi, S.Q.A. (2003) Additives and additive chemistry, in *Fuels and Lubricants Handbook: Technology, Properties, Performance and Testing* (eds G.E. Totten, R.J. Shah and S.R. Westbrook), ASTM.
[7] Bhushan, B. (1999) *Principles and Applications of Tribology*, John Wiley & Sons, Canada.
[8] Kenbeek, D. and Bünemann, T.F. (2003) Organic friction modifiers, in *Lubricant Additives: Chemistry and Applications* (ed. L.R. Rudnick), Marcel Dekker Inc., New York.
[9] Farng, L.O. (2003) Ashless antiwear and extreme pressure additives, in *Lubricant Additives: Chemistry and Applications* (ed. L.R. Rudnick), Marcel Dekker Inc., New York.
[10] Phillips, W.D. (2003) Ashless phosphorus containing lubricating oil additives, in *Lubricant Additives: Chemistry and Applications* (ed. L.R. Rudnick), Marcel Dekker, Inc., New York.
[11] Heuberger, R.C. (2007) *Combinatorial Study of the Tribochemistry on Anti-Wear Lubricant Additives*, ETH, Zurich.
[12] Covitch, M.J. (2003) Olefin copolymer viscosity modifiers, in *Lubricant Additives: Chemistry and Applications* (ed. L.R. Rudnick), Marcel Dekker Inc., New York.
[13] Kinker, B.G. (2003) Polymethacrylate viscosity modifiers, in *Lubricant Additives: Chemistry and Applications* (ed. L.R. Rudnick), Marcel Dekker Inc., New York.
[14] Zhang, J., Wu, C., Li, W. *et al.* (2003) Study on performance mechanism of pour point depressants with differential scanning calorimeter and X-ray diffraction methods. *Fuel*, **82**(11), 1419–1423.

[15] Rizvi, S.Q.A. (2003) Dispersants, in *Lubricant Additives: Chemistry and Applications* (ed. L.R. Rudnick), Marcel Dekker Inc., New York.
[16] Rizvi, S.Q.A. (2003) Detergents, in *Lubricant Additives: Chemistry and Applications* (ed. L.R. Rudnick), Marcel Dekker Inc., New York.
[17] Migdal, C.A. (2003) Antioxidants, in *Lubricant Additives: Chemistry and Applications* (ed. L.R. Rudnick), Marcel Dekker Inc., New York.

Part Two

Lubricant Performance

5
Formulating Lubricants

This chapter, being the first in the second part of the book, will cover the performance of lubricants, starting with formulation. Formulation involves analysing the application requirements, considering how to fulfil them with a good combination of base fluids and additives, and analysing the properties of the formulation.

This chapter describes general aspects on formulating lubricants and how to obtain and maintain high quality lubricated tribological contacts. In order to understand the idea of formulations, applications of hydraulics, gears and combustion engines are also described. The selected applications will also be used as examples in the following Chapters 6, 7 and 8.

5.1 General Aspects of Development

Formulating a lubricant requires application knowledge, which is converted into a mixture of base fluids and additives. There is a wide range of base fluids to select from and the choice of base fluids will indicate the type of lubricant that will be formulated. The application requirements as well as the positioning of the product will govern the selection of base fluids. The amount and type of additives added will depend on the lubricant type.

This section will cover formulations in general, development work, material compatibility, miscibility and interactions in a lubricated contact. All are important parts to consider for high quality products.

5.1.1 Formulations

Lubricants are a mixture of base fluids and additives (see Figure 5.1). The selection of base fluids is the first and key step in the formulation work. The lubricants will in most cases consist of mostly base fluids. Thus, the properties of the base fluid will serve as the basis for the lubricant properties, such as inherent viscosity, density and oxidation stability.

The base fluids selected may be mineral base oils, such as paraffinic base oils, naphthenic base oils or white oils. Paraffinic base oils are commonly used for mainline products. These products have good quality and perform according to indicated standards. The volumes of

---

Lubricants: Introduction to Properties and Performance, First Edition.
Marika Torbacke, Åsa Kassman Rudolphi and Elisabet Kassfeldt.
© 2014 John Wiley & Sons, Ltd. Published 2014 by John Wiley & Sons, Ltd.

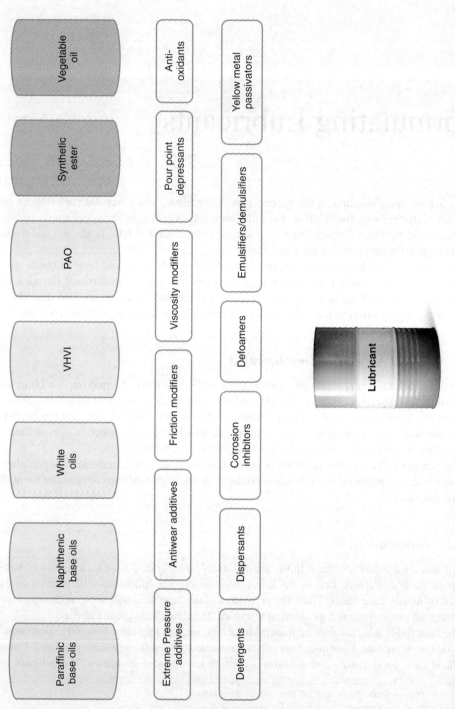

Figure 5.1 Lubricants can be formulated with a variety of base oils and additives (For a colour version of this figure, see the colour plate section)

naphthenic base oils are lower on the world market and they are used when, for example, good low temperature properties are desired. White oils can be selected in order to obtain a product with good working environment properties. Synthetics base oils, such as VHVI and PAO, are selected to obtain, for example, higher oxidation stability. These products are launched as high performance premium products with long life. Products reducing the environmental impact are formulated with either natural (i.e. vegetable oils) or synthetic esters. These allow for higher biodegradability and renewability than regular mineral base oils.

A lubricant may consist of one or several base fluids. One reason for having more than one base fluid in the formulation is to adjust the viscosity of the final lubricant. Other reasons for having several base fluids are to allow the lubricant to benefit from the properties of several base fluids or to optimize the formulation cost.

Adding additives to the base fluids will enhance the properties of the final lubricant. One or several additives will be added to the base fluids depending on the requirements of the product. Sets of different additives are commonly arranged in packages including all necessary additives or only some of them. The base fluid mainly governs the performance properties. However, viscosity modifiers and pour point depressants may be added to enhance these properties. Adding antioxidants, detergents, dispersants, defoamers and emulsifiers can enhance long life properties (see Figure 5.2).

The quality of the lubricated tribological contact can also be improved by adding additives. For example, corrosion inhibitors and yellow metal passivators will prevent surface degradation. For contacts working in the boundary lubrication and the mixed lubrication regimes, extreme pressure additives, antiwear additives and friction modifiers can be added to reduce friction and wear and thus prolong the life of the contacting surfaces.

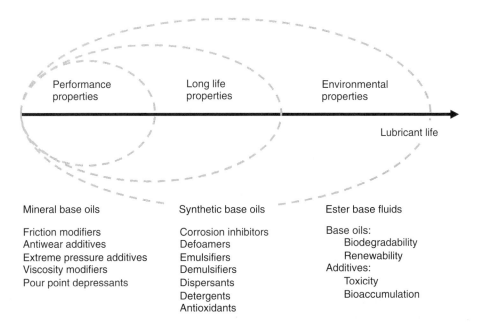

Figure 5.2 The lubricants affected by the base fluids, the additives and their environmental properties

Environmental properties are affected by the selection of base fluids such as white oils for working environment purposes or esters for biodegradability. Also, the additive chemistry can be monitored and the selection of, for example, low toxic and nonaccumulative additives will allow a product with a reduced impact on the environment to be formulated.

In general, base fluids are bought in large volumes for a relatively low price. Therefore, the amount of base fluids in the lubricant will in most cases not be limited by the cost of the product. Instead, as additives are more expensive than the base fluids they are added to meet the requirements only and not to exceed them.

5.1.2 Development Work

Lubricant development involves several overall common steps, such as having a requirement specification, blending in laboratory scale, analysing and testing.

It is crucial to understand the requirements before the development work begins. The requirements may come from a customer when developing a customer specific product or be stated in a standard when developing a product available for all customers. At this stage it is important to understand how the product will be marketed, that is core product (mainline or premium), customer specific or market specific. The agreed upon requirements for the product will serve as the basis for the development work and the properties of the developed product.

The development plan is done when the requirements are understood. The plan will include time, cost and the intended route for development. Analyses will be carried out, most likely several times. This work is repetitive in its character in one sense, but difficult to predict in another. The developer commonly runs into new problems along the way, making the development work more difficult and certainly interesting.

Examples of different types of oil analyses being run can be studied in more detail in Chapter 2 and examples of testing will be covered in Chapter 7.

5.1.3 Material Compatibility

A lubricant will most likely be brought into contact with several different materials in the real application. These materials include, for example, different metals, polymeric materials, elastomers, ceramics, leather or even wood. Both the material and the lubricant may be affected by the interaction. Therefore, it is crucial to ensure compatibility to secure a proper application function. During development, material compatibility is evaluated.

The compatibility between lubricants and metals is carried out as corrosion tests at elevated temperatures, which was described in Chapter 2. The evaluation is done visually where the severity of the corrosion of the metal is graded. However, by using surface characterization methods, as described in Chapter 8, a better understanding of the severity and the mechanisms can be revealed.

Polymeric materials comprise plastic and elastomeric materials. Plastic materials are found as packaging materials (i.e. the bottle containing the lubricant), hoses or in other details. They are commonly evaluated by bringing the material in contact with the lubricant at the application temperature, or at higher temperature, for a certain time period. The plastic material is visually inspected after testing and, for example, hardness and brittleness are compared before and after testing.

Elastomers are used for examples in seals to avoid leakage of the lubricant out of the system or entrainment of, for example, air and water into the lubricant. Elastomers are diverse and have traditionally been adapted to mineral base oils. Thus, using the most common elastomers such as NBR (nitrile butadiene rubber), HNBR (hydrogenated nitrile butadiene rubber) or FKM (fluoro rubber) in applications with a lubricant based on mineral base oils will in most cases work well. However, compatibility testing may be required when the lubricant is formulated with base fluids other than mineral base oils. Testing involves bringing the elastomer in contact with the lubricant at the application temperature or at an elevated temperature for a certain time period. Properties examined during testing comprise volume change, hardness change, tensile strength and elongation before break [1].

Ceramics, leather and wood are less common materials in a lubricated contact, although some low friction coatings can be classified as ceramics. When such materials are used in an application, the developer should pay attention to this and test the lubricant compatibility for the material of relevance.

5.1.4   Miscibility

Miscibility of one lubricant with another may be relevant during development when a new product is intended to replace an existing product. Whenever a change is done it is important to consider miscibility of the two lubricants.

Mixing two lubricants may affect, for example, the lubrication of the contact. Poor miscibility may affect air release, foaming, demulsibility and filterability. Therefore, tests are performed to evaluate the consequences of

- a top-up, where new lubricant is added to an already filled system,
- a 50–50% mix with old and new lubricant and
- filling the system with a new lubricant without emptying and cleaning it.

Even small amounts of surface active components may ruin miscibility. Therefore, tests proving miscibility are important to ensure a safe changeover to a new lubricant in the application.

5.1.5   Interactions in a Lubricated Contact

The lubricant will interact with different surfaces such as solid, liquid and gaseous surfaces when it is in use in the application, as has previously been discussed in Chapter 4. Solid surfaces may be, for example, metallic, polymer or ceramic materials. Liquid surfaces may be, for example, different base fluids or water entrainment. Gaseous surfaces could be, for example, foam or the free surface on top of a container or blender. During the formulation work this is taken into consideration as far as possible.

A lubricant is a sensitive balance of base fluid and additive molecules, where the additives sometimes counteract and sometimes enhance the action of each other. Consequently, the formulation handles interactions between base fluid molecules and additive molecules, as well as between base fluid molecules, additive molecules and surfaces. The interactions may be weak or strong, depending on the different molecular chemistries (see Figure 5.3).

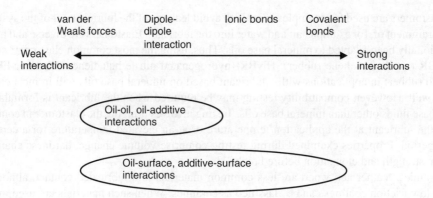

Figure 5.3 Different possible interactions between base fluids, additives and surfaces in the lubricated contact

In this context, the terms 'nonpolar base fluids' and 'polar base fluids' are relevant (refer to Chapter 4 for more details). Nonpolar base fluids have an even distribution of electrons within the base fluid molecule. This is the case for mineral base oils. Base fluids with oxygen in the molecular structure will have an asymmetrical distribution of electrons towards the electronegative oxygen, making the base fluid polar. Natural and synthetic esters are polar base fluids. Permanent dipole moments exist in some of these base fluids due to an asymmetrical arrangement of positively and negatively charged parts of the base fluid molecule.

The interactions are in this context described by van der Waals forces, dipole–dipole interactions, covalent bonds and ionic bonds. In a lubricant where no chemical reactions occur, only van der Waals forces and dipole–dipole interactions exist. Thus, mineral base oil molecules interact with each other or nonpolar additives with van der Waal forces.

Dipole–dipole interactions occur when either the base fluid or additive molecules are polar or between polar base fluids, polar additives and metal surfaces. Several additives or ester base fluids are surface active (i.e. polar). Some additives are tailored to interact with surfaces such as friction modifiers. Others are tailored to interact more strongly with surfaces, forming covalent or ionic bonds, that is resulting in a chemical reaction. An example is when extreme pressure additives react with a steel surface forming a low friction layer of iron sulfide.

In order to be able to market a functioning lubricant for the intended application, all these interaction phenomena need to be considered, understood, tested and solved during development. This is important for the lubricant to operate well in the tribological contact.

5.2 Quality of the Lubricated Tribological Contact

The main aspect when formulating or selecting a lubricant for a specific tribological contact or component is to obtain a contact situation where a good function of the component is maintained for a long service time. Thus, the quality of the lubricated contact includes both short term and long term aspects. In this section the procedure to obtain and maintain a high quality contact is discussed.

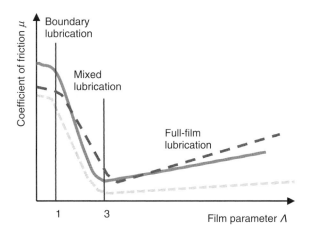

Figure 5.4 Two examples of lubricants showing better performance than the reference lubricant (For a colour version of this figure, see the colour plate section)

5.2.1 Lubricant Film Regime

Friction is an important aspect of the contact quality. As has been previously described (refer to Figure 1.10), contacts can operate in different lubrication regimes. The best lubricant for a specific tribological contact could, for example, be the one giving the easiest transition to full-film lubrication (e.g. the lubricant giving full-film lubrication at the lowest speed and/or highest load), or the lowest friction in the full-film regime, or the one giving the lowest friction in the boundary lubrication regime (see Figure 5.4).

Lubricants are formulated with a desired quality when new, and the formulation also includes additives to prevent ageing. However, oxidation, corrosion and contamination may still reduce the quality of the lubricant. Also, wear and corrosion may degrade the quality of the surfaces in contact. Consequently, the first step towards a high quality lubricated contact is to guarantee that the required lubricant film regime and friction level are obtained with the new oil. The next step is to guarantee that good contact conditions are maintained for a long service life.

To maintain good contact conditions requires an understanding of different contact parameters and processes (see Figure 5.5). These will therefore be covered here in more detail.

The lubricant film regime is determined by the lubricant film thickness and surface roughness, as expressed in Equation (1.5). The relation shows that reducing the surface roughness gives a larger film parameter Λ. This implies that having smoother surfaces may result in a transition from the mixed to the full-film lubrication regime.

The lubricant film thickness depends on the design and function of the component, including lubrication. It is given by the mechanical parameters in combination with lubrication parameters.

Mechanical parameters affecting lubricant film thickness are load and pressure distribution, type of motion and velocity. The total load and pressure distribution in the contact depends on the contact geometry, the alignment in the contact, the applied load and the material properties. The type of motion and velocity is given by the function of the component.

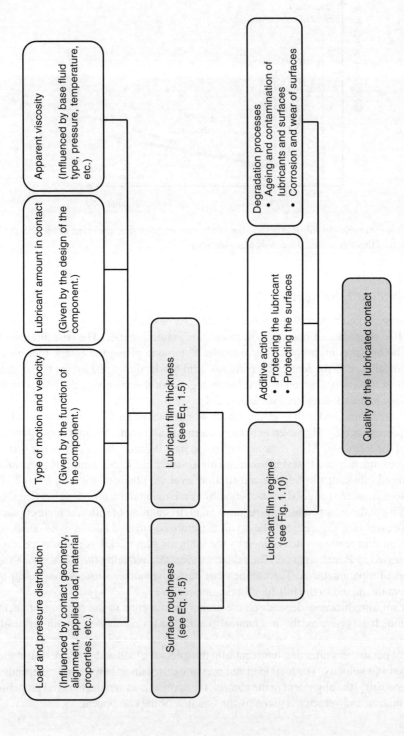

Figure 5.5 Parameters affecting the final lubricant quality and the film regime

Lubrication parameters are the amount of lubricant in the contact and its apparent viscosity (refer to Chapters 1 and 2). The apparent viscosity is the viscosity that results at a certain pressure and temperature for the base fluid used, that is the viscosity under the actual operating conditions. Thus, different base fluids may have different apparent viscosities at the same operating pressures and temperatures. Also, the apparent viscosity for different base fluids shows a different dependency of pressure and temperature.

5.2.2 Maintaining a High Quality Contact

If the lubricant and the metal surfaces in the tribological contact interact well the high quality of the contact will be maintained. However, during operation the conditions may change, for example the temperature may rise or some of the additives may be consumed.

For contacts operating in the full-film regime a change of the lubricant viscosity may, for example, increase the friction or lead to a transition to the mixed-film regime. The change in viscosity may be due to high shearing, high temperature, oxidation or contamination of the lubricant. Thus, for lubricants intended to operate in the full-film regime it is crucial to monitor the viscosity and take necessary precautions if changes occur. It may result in a change of lubricants in the system. The reason for the viscosity change needs to be investigated, avoiding future viscosity changes.

In the mixed-film regime the contacting surfaces may start to wear or be exposed to fatigue. For contacts operating in the boundary or mixed-film regimes it is important that the contacting surfaces are protected from high friction, wear and corrosion. Thus, the function of the additives is important. For lubricants intended to operate in the mixed or boundary lubrication regime it is important to regularly analyse the lubricant to ensure that proper additives are present in the right amounts. If the additive levels decrease too much it will be necessary to replace the lubricant to avoid wear and in worst cases surface fatigue.

5.3 Hydraulics

Hydraulics is used for generation, control, and transmission of power by the use of pressurized fluids. Hydraulic systems occur in for example, construction equipment, steel plants, machine tools, marine applications, and in the mining industry. The main purpose of hydraulics is to transfer power from one point to another. They can be built in small units with high power density [2, 3].

5.3.1 Description of a Hydraulic System

An example of a hydraulic system is shown in Figure 5.6. In this case, the hydraulic oil transfers power from the pump to the piston or engine. The hydraulic oil is stored in a tank from where it is pumped through valves. The system is a closed loop and to increase system cleanliness the oil is filtered before entering the tank again. The valves may be monitored to control the hydraulic oil to be pumped to the piston or engine, or back to the tank via the filter.

The demands on hydraulic systems are high and steadily increasing. They include higher performance (i.e. pressures), higher degree of cleanliness, extended drain intervals and increased environmental demands, in combination with reduced oil volumes. The environment may

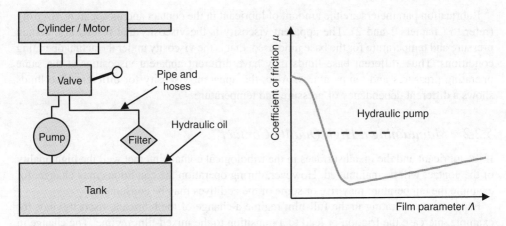

Figure 5.6 A hydraulic system is visualized (left) and the lubricating conditions are shown (right)

be full of contaminants, such as particles or water. Regardless of a harsh environment, the hydraulic system should operate as intended. Hydraulic systems operate under minimum friction covering full-film lubrication to mixed lubrication (see Figure 5.6). A hydraulic system should prevent wear and corrosion, remove heat and operate under clean conditions. This can be performed through optimum design together with the action of the lubricant. The performance of hydraulic oils is governed by standards and specifications, setting the performance limits for the application, including the lubricant. Hydraulic oils that fulfil specifications are placed on certain official lists, which are available for consumers to use. These lists serve as a guarantee for the lubricant properties and performance to customers.

There are hydraulic systems for indoor applications running in fairly clean environments at moderate temperatures. The lubricant life of indoor hydraulic oils is fairly long. Outdoor applications are exposed to a broader temperature range, allowing operability at both moderate and very low temperatures. Also, many outdoor hydraulic systems are mobile, with a high risk of leakage into sensitive environments. In some cases the hydraulic system should promote a clean working environment or operate at high temperatures without catching fire.

5.3.2 Formulating Hydraulic Oils

Hydraulic oils are used in hydraulic systems, as their names indicate, and should therefore fulfil the requirements for the full hydraulic application. Hydraulic oils are primarily described by their viscosity and are formulated to have viscosities between 15 and 100 cSt at 40 °C [4,5]. The viscosity is described by viscosity grades ISO VG[1] (see Table 5.1).

Second, hydraulic oils are described according to the base fluids used. Different base fluids are used depending on the application. Most hydraulic oils are formulated with mineral base oils allowing for a mainline product with good price performance. White oils are used in

---

[1] International Standards Organisation

Table 5.1 The viscosity classification according to ISO shown as ISO VG (i.e. viscosity grades; KV_{40}, kinematic viscosity at 40 °C)

ISO VG	KV_{40} (cSt)	KV_{40}, minimum (cSt)	KV_{40}, maximum (cSt)
15	15	13.5	16.5
22	22	19.8	24.2
32	32	29.8	35.2
46	46	41.4	50.6
68	68	61.2	74.8
100	100	90.0	110
150	150	135	165
220	220	198	242
320	320	288	352
460	460	414	506
680	680	612	748
1000	1000	900	1100

hydraulic oils for work–environment reasons, commonly indoor hydraulic systems. Synthetic esters or vegetable oils are used in environmentally adapted hydraulic oils exposed to a higher risk of leakage, usually outdoor applications (e.g. garbage trucks).

Other factors are determined via specifications such as ISO 15380 (mentioned in Section 2.3), which covers requirements for hydraulic oils based on vegetable oils, polyalkyleneglycols, synthetic esters and PAOs. Properties governing the quality needed are stated. Some properties are not pinpointed with a value, but the supplier of the lubricant needs to report the value or need to agree with the customer on a suitable level.

There are some requirements that are specific to hydraulic applications. These primarily involve pump tests for wear testing and low temperature viscosity tests. Pump tests will evaluate the antiwear properties of hydraulic oils and low temperature viscosity tests will evaluate the viscosity modifier content of outdoor hydraulic oils. The specifications may further involve temperature-related properties, such as flash point and pour point. Both are important to secure functionality under the given operational conditions. Also, air release and foaming are given limits in the specification. It is important to have low air release values to avoid cavitation in the hydraulic pump and low foaming to minimize foam build-up in the system. Corrosion inhibition will ensure a longer lubricant and metal surface life.

The standard also specifies material compatibility with, for example, elastomers (i.e. sealing materials) and metallic materials. The hydraulic lubricant should be formulated to be able to interact well with the elastomer materials commonly used, where hardness change, volume change, elongation and tensile strength need to be in accordance with set limits. It is further recommended to avoid pure lead, tin and zinc due to the risk of soft metal corrosion, particularly at elevated temperatures and in contact with aged lubricants.

Change intervals of hydraulic lubricants can be based on experience, by topping up regularly with new lubricant or monitored via analyses on a regular basis. The changing of the lubricant will be discussed in Chapter 7. The changing of filters, which is mentioned in the standard ISO 15380, is considered part of the application knowledge and will not be covered here.

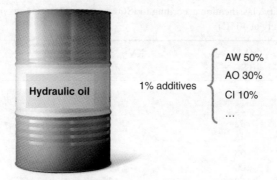

Figure 5.7 General formulation description of indoor hydraulic oils. Outdoor hydraulic oils contain viscosity modifiers in addition to the shown additives

Environmental property requirements are important for a sustainable product. They will be more important for a product where there is a risk of leakage to the environment or when there is a risk of exposure to operators. In ISO 15380, the environmental properties stated include biodegradability and toxicity for fish, daphnia and bacteria.

In order to fulfil the criteria indicated in, for example, ISO 15380, hydraulic oils for indoor applications consist of mainly base fluids with less than 1% additives. Hydraulic applications require low wear and long life. Therefore, hydraulic oils contain relatively high amounts of antiwear additives and antioxidants (see Figure 5.7). Hydraulic oils for outdoor applications require viscosity modifiers in order to operate at the wide temperature range required. Therefore, the amount of additives is higher for outdoor hydraulic oils.

5.4 Gears

Gears are rotating machine parts with teeth meshing another toothed part in order to transmit torque (see Figure 5.8). They occur in both industrial and automotive applications. Industrial gears are used, for example, in steel mills, mining, power generation, agriculture, manufacturing and forestry. There are several gear types used, such as spur gears, helical gears, bevel gears and worm gears.

5.4.1 Description of Gears

Different types of gears have varying degree of sliding and rolling motion in the contact. The load may be moderate to very high and many gears operate in the EHD regime. However, gears also operate in the boundary lubrication regime during start-up and stop [6, 7].

An industrial gear lubricant should form a thick enough lubricating film to separate the surfaces and thus prevent wear and reduce friction. It should further remove heat, lubricate bearings and prevent corrosion. The demands on industrial gear oils are continuously increasing with increasing power output. The demands on the gears continuously increase with slow moving gears and higher power outputs in combination with decreasing sizes. The optimum operability of gears is allowed by optimum design together with a balanced fully formulated gear lubricant. The performance of gear lubricants is stated in gear standards.

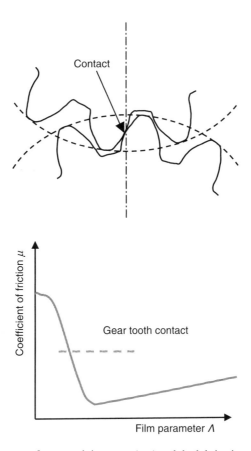

Figure 5.8 The contact of gear teeth in a gear (top) and the lubricating conditions (bottom)

5.4.2 Formulating Gear Oils

Industrial gear oils are classified according to the ISO VG, as is the case for the hydraulic oils. Gear oils have higher viscosity grades than hydraulic oils and are formulated from 68 to 1000 cSt at 40 °C (see Table 5.1). Automotive gear oils are classified according to SAE[2] J306 (see Table 5.2).

Gear oils are commonly formulated with mineral and synthetic base fluids, which are stable and offer good price performance [8]. They should primarily prevent scuffing and wear. In addition, they should withstand high local temperatures, reduce micropitting and reduce oxidation and corrosion.

The properties are stated in gear specifications. One common standard for reference is DIN 51517, where the requirements of gear oils with different amounts of additives are specified. There are some gear specific tests in the standard, such as gear tests to ensure that the gear oil has a good ability to carry high loads in the gear tooth contact. In addition to carry load,

---

[2] Society of Automotive Engineers

Table 5.2 Viscosity classification of automotive gears according to SAE J306 (KV_{100}, kinematic viscosity at 100 °C)

SAE viscosity grade	Maximum temperature for viscosity of 150 000 cP (°C)	KV_{100}, minimum (cSt)	KV_{100}, maximum (cSt)
70W	−55	4.1	—
75W	−40	4.1	—
80W	−26	7.0	—
85W	−12	11.0	—
80	—	7.0	<11.0
85	—	11.0	<13.5
90	—	13.5	<18.5
110	—	18.5	<24.0
140	—	24.0	<32.5
190	—	32.5	<41.0
250	—	41.0	—

the gear oils should be able to perform at different working conditions, giving requirements for viscosity, flash point and pour point. The standard also specifies long life properties, such as foaming, demulsibility, steel and yellow metal corrosion and oxidation stability. Also, compatibility with elastomers is stated in the same manner as it is for hydraulic oils.

Gear oils require higher amounts of additives than hydraulic oils (see Figure 5.9). The main difference in formulation is the addition of EP additives for minimizing wear during the high loads in boundary lubrication.

Manufacturers test gear oils in their own gears to guarantee the functionality of different brands of gear oils. These tests are performed in bench tests at the manufacturers' sites. Thus, it is not enough to fulfil the gear specifications. In addition to meeting the specifications, gear tests have to be carried out in most cases.

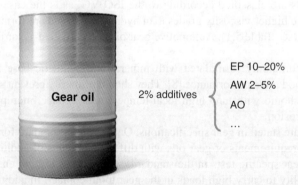

Figure 5.9 General formulation description of industrial gear oils

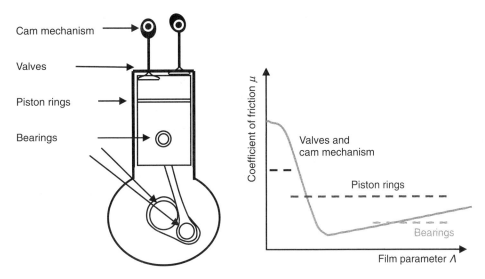

Figure 5.10 The lubricated parts in an engine (left) and the corresponding lubricating conditions indicated in the diagram (right)

5.5 Combustion Engines

Engines convert energy into useful mechanical motion. Commonly the energy is retrieved from the combustion of fuels. Combustion engines are found in different sizes and in a variety of vehicles. However, the operation is similar regardless of engine size [9].

5.5.1 Description of Combustion Engines

The engine operation will be described from the four-stroke engine perspective (see Figure 5.10). During intake, air is sucked into the combustion chamber. Fuel is injected during the compression stroke. Different fuels are used for different engines. Fuels generally consist of carbon and hydrogen, C_xH_y. They may include gasoline (having 5–12 carbons), diesel or biodiesel (having 10–15 carbons), ethanol or gaseous fuels (e.g. natural gas, biogas or landfill gas). Fuel and air react during the compression stroke.[3] The combustion energy released pushes the piston downwards, yielding power, heat and friction during the expansion stroke.

The final stroke releases the exhaust components.[4] The current emission legislation (i.e. Euro V) has limits on the allowable exhaust components (e.g. CO, hydrocarbons, NO_x and

---

[3] Combustion results in different products depending on the combustion temperature. At low temperatures CO_2 and H_2O form. At high combustion temperatures CO_2, H_2O and NO_x form.

[4] These vary depending on the conditions in the combustion chamber. The exhaust consists of CO_2 and H_2O during lean fuel conditions. Under fuel-rich conditions the exhaust may consist of CO_2, H_2O, CO, H_2, unburned hydrocarbons and soot.

particulate matters, i.e. soot). Emissions may be reduced by after-treatment of the exhaust gas or optimizing the combustion.[5]

In this section passenger car engines, heavy duty engines, marine diesel engines and gas engines are considered to show the variety of the applications. *Passenger car engines* are constructed to operate for 250 000–400 000 km. They should run smoothly regardless of season. They can be run on gasoline, diesel or ethanol fuels [10]. *Heavy duty engines* are found in commercial vehicles such as trucks. They are constructed to last for 700 000–800 000 km. They are expected to run smoothly under high loads regardless of the surrounding environment. Heavy duty engines are run on diesel/biodiesel fuel.

Marine diesel engines are used in, for example, ferries and passenger ships. They are also used in larger tankers and bulk carriers. They are expected to last for a very long time, such as 20–25 years at 24-hour operation. Marine diesel engines may be run on diesel or bunker fuel (heavy fuel oil) with up to 5.5% sulfur. *Gas engines* are used primarily for generators and pumps in gas pipelines. Today, there is increased usage in marine applications. They are large engines that are expected to last for 10 years of operation.

The whole engine uses one single lubricant, which should lubricate, for instance, the piston ring-cylinder, the bearings and the valve train, as well as cool the pistons. The piston ring-cylinder operates in the boundary regime at the turning points and reaches full-film lubrication in between the turning points. The rotating engine bearings are full-film lubricated. The valve train with the cam and follower interface is in the boundary and the mixed lubrication regimes.

5.5.2 Formulating Combustion Engine Oils

The demands on extended drain, fuel economy and emission limits have become so strict that engine oil formulation significantly contributes to overall engine performance. The lubricant must keep the engine clean in order to allow for extended drain intervals. This is done by adding dispersants and detergents to the engine oil.

The engine oil should lubricate, for instance, the bearings, the cylinders, the camshaft and the valve train, as well as cool the pistons. Wear protection is secured by antiwear additives. The engine oil should lubricate the engine both at low and high temperatures. This is performed by adding viscosity modifiers, pour point depressants and antioxidants to the engine oil.

Emission limits have placed chemical limits on the engine oils. Engine oils with chemical limits are called low (or mid) SAPS. These limits comprise sulfated ash (SA), phosphorus (P) and sulfur (S), since these chemicals have been shown to affect after-treatment devices such as filters and catalysts negatively. Therefore, chemical limits have been determined to prolong the life of the after-treatment devices.

Engine oils are heavily regulated via specifications and requirements from original equipment manufacturers (OEMs). The OEMs require certain tests to be passed to fulfil guarantee obligations. Engine oil development is extremely costly due to the tough testing requirements.

---

[5] Catalysts to remove NO_x, filters to remove particulate matters, improved fuel economy to lower CO_2 emissions and improved combustion to reduce NO_x and minimize hydrocarbons and soot. The combustion can be improved by optimizing the design of the fuel injectors, optimizing the fuel-to-air ratio and controlling the temperature in the combustion chamber.

Table 5.3 The viscosity classification of engine oils according to SAE J300. W (as in winter) is used for low temperatures

SAE viscosity grade	CCS max (mPa s) at °C	MRV max (mPa s) at °C	KV_{100} min (mm²/s) at 100 °C	HTHS min (mPa s) at 150 °C
0W	6200 at −35	60 000 at −40	3.8	—
5W	6600 at −30	60 000 at −35	3.8	—
10W	7000 at −25	60 000 at −30	4.1	—
15W	7000 at −20	60 000 at −25	5.6	—
20W	9500 at −15	60 000 at −20	5.6	—
25W	13 000 at −10	60 000 at −15	9.3	—
20	—	—	5.6 (max < 9.3)	2.6
30	—	—	9.3 (max < 12.5)	2.9
40	—	—	12.5 (max <16.3)	3.5
50	—	—	16.3 (max <21.9)	3.7
60	—	—	21.9 (max < 26.1)	3.7

CCS = cold crank simulator (dynamic viscosity at low temperature), MRV = mini-rotary viscometer (dynamic viscosity for low temperature pumping), KV_{100} = kinematic viscosity at 100 °C, low shear and HTHS = high temperature, high shear (dynamic viscosity)

However, the engine oils have a high quality, which is ensured by the OEM approvals and the ACEA[6] specifications on the container. No field tests are required for formal approvals.

Engine oils operate under a wide range of temperatures and a wide range of shear rates. They are therefore classified according to viscosities at low and high temperatures. This implies diverse viscosity measurements for engine oil applications. Common viscosity measurements for engine oils are: CCS = cold crank simulator (dynamic viscosity), MRV = mini-rotary viscometer (dynamic viscosity), HTHS = high temperature, high shear (dynamic viscosity) and KV_{100} = kinematic viscosity at 100 °C. These tests will ensure that the combustion engine starts even if it is very cold, secure pumpability of the lubricant at low temperatures and guarantee good fuel economy.

Engine oils are specified according to SAE J300 (see Table 5.3) [11]. This table can be used for both monograde and multigrade lubricants. Monograde lubricants fulfil either the cold (W for winter) or the warm part of the table. Multigrade lubricants fulfil both cold and warm tests in the table. For instance, engine oils may be branded as 5W-40, meaning that it fulfils the viscosity requirements both on the cold side and the warm side. Thus, if the minimum requirements differ the highest minimum value is relevant for a multigrade lubricant.

Mineral engine oils are formulated with paraffinic mineral base oils, resulting in a premium product with a good price performance. VHVI or PAO base oils are used in long life synthetic combustion engine oils. Sometimes mineral base oils are mixed with synthetic base oils. This is done in semi-synthetic engine oils.

---

[6] ACEA = Association des Constructeurs Européens d'Automobiles, or the Association of European Automobile Manufacturers

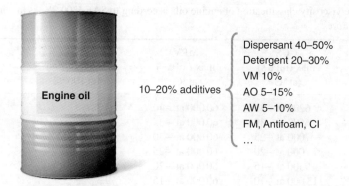

Figure 5.11 General formulation description of combustion engine oils

Specifications in Europe are set by ACEA. The main physical and chemical requirements are HTHS viscosity, sulfated ash, phosphorus and sulfur limits, evaporation loss and TBN values. In addition to these tests, several bench tests are needed during development, ensuring that required levels of, for example, sludge formation, piston cleanliness and ring sticking are met. Bench tests will be further covered in Chapter 6.

Engine cleanliness is important for engine oils in order to meet oil change intervals. Adding dispersants and detergents ensures engine cleanliness. The amount of additives in engine oils varies between 10 and 20%, with mostly dispersants and detergents (see Figure 5.11).

The formulations differ between engine oils for passenger cars, heavy duty vehicles, marine diesel engines and gas engines. The main difference is the chemistry of the detergents and the dispersants as well as the treat rate of these additives.

For passenger car engines and heavy duty vehicle engines, the level of detergents and dispersants are high enough to guarantee the functionality of the engine at the required change interval. Commonly the TBN value is around 10 in both types of engines. However, the detergents and dispersants in heavy duty engine oils are more powerful than the ones used in passenger car engine oils.

The detergents and dispersants in marine diesel engine oils should be able to handle combustion residues from fuels containing as much sulfur as 5.5%. The sulfur may react, forming acids in the engine oil. These acids may give rise to corrosion unless they are neutralized by the detergent chemistry. The levels of detergents and dispersants are very high and the TBN numbers may reach 40.

Gas engines are run on biogas, natural gas or landfill gas. Depending on the gas used, different detergents and dispersants are used at different treat rates. If the gas engine is run on natural gas, which is a very clean fuel, the TBN number can be as low as 0–6. Both biogas and landfill gases are chemically less defined and give rise to higher demands on cleanliness in the lubricant. Therefore, TBN numbers of 10 are common in gas engine oils used for these fuels.

References

[1] Torbacke, M. and Johansson, A. (2005) Seal material and base fluid compatibility: an overview. *Journal of Synthetic Lubrication*, **22**, 2.

[2] Tessman, R.K., Melief, H.M. and Bishop, R.J. (2006) Basic hydraulic pump and circuit design, in *Handbook of Lubrication and Tribology, Vol. 1: Application and Maintenance*, 2nd edn (ed. G.E. Totten), Taylor & Francis.
[3] Murrenhoff, H., Göhler, O.-C. and Meindorf, T. (2006) Hydraulic fluids, in *Handbook of Lubrication and Tribology, Vol. 1: Application and Maintenance*, 2nd edn (ed. G.E. Totten), Taylor & Francis.
[4] Murrenhoff, H. and Remmelmann, A. (2003) Environmentally friendly oils, in *Fuels and Lubricants Handbook: Technology, Properties, Performance and Testing* (ed. G.E. Totten), ASTM International.
[5] Givens, W.A. and Michael, P.W. (2003) Hydraulic fluids, in *Fuels and Lubricants Handbook: Technology, Properties, Performance and Testing* (ed. G.E. Totten), ASTM International.
[6] Tipton, C.D. (2006) Automatic transmission fluids, in *Handbook of Lubrication and Tribology, Vol. 1: Application and Maintenance*, 2nd edn (ed. G.E. Totten), Taylor & Francis.
[7] Gangopadhyay, A. and Qureshi, F. (2006) Rear axle lubrication, in *Handbook of Lubrication and Tribology, Vol. 1: Application and Maintenance*, 2nd edn (ed. G.E. Totten), Taylor & Francis.
[8] Bala, V. (2003) Gear lubricants, in *Fuels and Lubricants Handbook: Technology, Properties, Performance and Testing* (ed. G.E. Totten), ASTM International.
[9] Heywood, J.B. (1988) *Internal Combustion Engine Fundamentals*, McGraw-Hill.
[10] Tung, S.C., McMillan, M.L., Becker, E.P. and Schwartz, S.E. (2006) Automotive engine oils, in *Handbook of Lubrication and Tribology, Vol. 1: Application and Maintenance*, 2nd edn (ed. G.E. Totten), Taylor & Francis.
[11] Schwartz, S.E., Tung, S.C. and McMillan, M.L. (2003) Automotive lubricants, in *Fuels and Lubricants Handbook: Technology, Properties, Performance and Testing* (ed. G.E. Totten), ASTM International.

6

Tribological Test Methods

There are many reasons for carrying out *tribological tests* or *tribotests*. One reason is to study the wear and friction mechanisms appearing in specific tribological applications. Other reasons are ranking of materials and lubricants for existing equipments or selection of materials and lubricants for new applications. Tribotesting may also be performed for general, application independent, characterization of wear and friction. Note that lubricant properties are tested according to the methods presented in Chapter 2, while the lubricant performance is evaluated by tribotesting.

Tribotests can be classified into tests that *simulate the function of real components* or tribological systems and tests that *simulate the critical tribological load*. The former class includes field tests, bench tests and component tests. The latter class comprises different model tests. All testing aims at increasing the tribological understanding at the fundamental or system level in order to enable development of design, construction and function of tribological systems. The complexity of testing may differ as well as the time and cost for testing (see Figure 6.1) [1–3].

Field tests are applied to evaluate tribological properties in a real application. They are generally run under ordinary operating conditions and the test duration is usually very long, that is weeks or months and in some cases more than a year. *Bench tests* are performed with parts or subsystems of a mechanical system or machinery in a laboratory environment. An example is testing of engine oils in real engines in a laboratory, rather than in a real vehicle as in a field test. *Component tests* are test set-ups designed with original parts, but in a simplified system. A typical component is bearings, and in a component test set-up several bearings can be tested simultaneously. *Model tests* are applied in order to simulate the critical tribological load of, for example, a component, a material, or a lubricant, in contrast to field, bench and component tests that rather simulate the function of real components. Model tests are performed in a laboratory environment under controlled and varying testing conditions.

6.1 Field, Bench and Component Tests

Field tests are applied to evaluate tribological properties in a real application. They are generally run under ordinary operating conditions and the test duration is usually very long, that is weeks

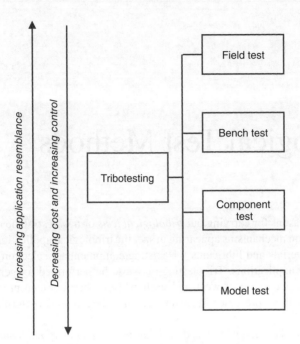

Figure 6.1 Different types of tribological tests

or months and in some cases more than a year. The outcome shows the functionality of the system. However, during field testing the tribological contact may be subjected to a variety of stress and load conditions. As a consequence, it may not be possible to evaluate the influence of different separate factors, such as water content in the lubricant, contact load variations and so on. In the case of a component or system failure, it may also be difficult to identify the exact reason for failure. Thus, field tests measure system properties without being able to characterize tribological contact properties. Fields tests are expensive due to the long testing period and the use of real components and conditions.

Bench tests are performed with parts or subsystems of a mechanical system or machinery in a laboratory environment. An example is testing of engine oils in real engines in a laboratory, rather than in a real vehicle as in a field test. In comparison with field tests the bench tests allow better control of the testing variables, such as fuel and oxygen consumption, as well as better possibilities for real time recording of test conditions, such as temperature, pressure or combustion products and emissions (e.g. CO_2 and NO_x). Since the contact conditions during the testing are fairly known, they can to some extent be correlated to the wear mechanisms of the surfaces of, for example, piston, cylinder and camshaft.

Gearboxes can also be tested in bench tests. These tests gives information on a system level, that is dynamic properties of the housing may influence the output. Tests of bearings mounted in original hub units can also be categorized as bench tests. The design of the hub may influence the performance of the core part, which is the bearing; therefore a test with the bearing and the hub mounted together is more realistic than using only a bearing.

Component tests are test set-ups designed with original parts, but in a simplified system. A typical component is bearings, and in a component test set-up several bearings can be tested simultaneously. Another example is testing of hydraulic cylinders. When testing a hydraulic cylinder the focus can be on the piston sealing or the piston rod sealing, but it is valuable to test the whole component. Component testing will, for example, allow lifetime estimation in any selected controlled environment, which is not possible in a field test. Control and recording of test conditions are possible. Component tests are less expensive than field tests and bench tests.

6.2 Model Tests

One reason for performing model tests may be to simulate the critical tribological load of a component in a real application. Another reason could be to carry out a more general characterization of the tribological properties of materials and lubricants, for example in order to map the possible range of use in terms of contact pressure and sliding speed.

Model tests are performed in a laboratory environment under controlled and varying testing conditions. This allows the evaluation of the testing to be very accurate. The results, including the evaluation of wear and friction mechanisms, can easily be correlated to the parameters varied during testing. Small size and simple geometry of test specimens make the tests relatively inexpensive, and several tests can often be performed in a short time period, which increase the accuracy of results.

Careful planning is required in order to perform model testing that simulates a real application. The model test selected should provide the closest possible resemblance to the application in mind. It is also important to make sure that the correct friction and wear mechanisms are imitated in the test. This means that the wear rate, surface topography, damage mechanisms, chemical composition, and so on, must be examined in order to verify that the same contact situation prevails during testing as for the real component in service. Finally, also for verification, appropriate reference materials and lubricants should be included in the testing. However, the transfer of model test results into real components or applications may be difficult and requires extensive experience and competence in the field of tribology.

For appropriate test planning, some characteristics of mechanical components can be pointed out: both mating surfaces are of equal importance, a relatively long life is expected, the surfaces are initially relatively rough and a running-in is necessary, the nominal contact pressure and the surface temperature are relatively low and a lubricant is present.

6.2.1 Strategy for Selecting and Planning a Model Test

The model test selected should provide the closest possible resemblance to the application in mind. The first step is to evaluate the contact geometry, that is the form or shape of the contacting bodies and whether the contact is conformal or nonconformal (see Figure 6.2) (also see Chapter 1 for more details). The contact geometry directly affects the local conditions in the contact and is considered to be the primary variable for selecting model test and for scaling up and scaling down of tests. Finally, the test duration must be set long enough for the test to be correctly evaluated.

The *contact geometry* will determine the size and distribution of the contact pressure. Conformal contacts have distributed surface areas, whereas nonconformal contacts may have line or point contact areas. The nominal contact pressure in a point or line contact can be

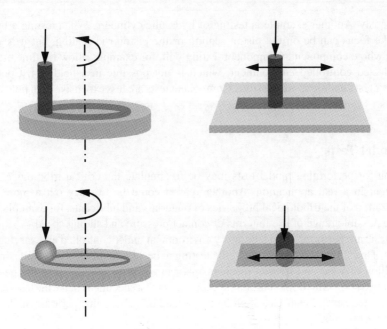

Figure 6.2 Different possible contact configurations for model tests evaluating sliding wear. To the left: pin-on-disc with uniform motion (top) and ball-on-disc with uniform motion (bottom). To the right: reciprocating pin against a fixed plate (top) and reciprocating cylinder against a fixed plate (bottom)

significantly higher (at least initially) than in a distributed contact area, since the applied load is spread over a much smaller area. With distributed contact areas the contact pressure may be constant throughout the test. However, alignment problems in the equipment may cause uneven wear across the surfaces, which is an important issue to deal with.

Then the *direction of motion* may be considered, stating whether the testing should be done in a unidirectional or in a reciprocating mode. In a unidirectional mode the relative motion at the contact interface has the same direction during the full test. Unidirectional contacts are found in, for example, cam–followers and journal bearings. In a reciprocating mode the motion repeatedly changes direction. This implies that the relative velocity is zero at the turning points where the direction changes. Reciprocating motion is found in, for example, piston–cylinder contacts where the piston reaches the end points and changes direction for each stroke.

Testing may be performed under *sliding* or *rolling* conditions. Also available are model tests where both solid surfaces may be rotated independently, allowing a combination of sliding and rolling. In such set-ups the sliding-to-rolling ratio can be selected and varied. In the contact between two gear teeth the relative velocity between the surfaces differs and there is a shift from a sliding and rolling situation to a pure rolling and further on to a sliding and rolling in the opposite direction.

When the contact geometry and the motion character have been determined, the *operating* and *environmental conditions* also have to be identified. Operating conditions include, for example, contact pressure and relative velocity. Environmental conditions include, for example, temperature and humidity. To simulate the application, similar conditions as in the real component should be possible to apply in the selected model test.

The aim of model testing may be to study the influence of variations in contact geometry, relative motion, operating conditions and environment. Here, *materials, surface treatments* and *lubricants* identical to those in the existing component should be used. This testing may identify the critical parameters for function and failure of the component. An alternative aim may be to study the use of new materials, surface treatments or lubricants. Here, it is important to include appropriate reference materials and lubricants. Generally, including the materials and lubricants that are used in the existing component is valuable for verification.

The *lubrication regime* also influences the selection of the model test. In the *full-film lubrication* regime the solid surfaces are fully separated by the lubricant. Since the solid surfaces are not in contact wear does not occur. Thus, results of interest from such model tests are film thickness and friction. *Elastohydrodynamic lubrication* is a special case of the full-film lubrication regime. The contact pressure is high enough to cause deformation of the solid surfaces and the lubricant solidifies in the contact. Thus, the evaluation of testing in this regime should include an analysis of changes of the contact geometry. In the *mixed lubrication* regime the solid surfaces are sometimes in contact with each other and in the *boundary lubrication* regime they are always in contact. Consequently, friction and wear are of interest to measure and evaluate, including evaluation of existence of tribofilms and films formed by the lubricant, while the lubricant film thickness is not relevant.

6.3 Lubricant Film Thickness Measurements

The lubricant film thickness provides information about the ability of the lubricant to build up a lubricating film in the contact under given contact conditions. It is important for the lubricant to build a thick enough film to perform well by separating the two solid surfaces. Designers strive to optimize the film thickness. If the film thickness is too low, wear will increase. If the film thickness is too high, the friction within the lubricant layers will increase, which implies unnecessary energy losses and heating of the lubricant.

A measurement of the film thickness is normally based on measurement of the gap between two solid surfaces. In hydrodynamic full-film lubrication it is assumed that the solid surfaces are not deformed. In elastohydrodynamic full-film lubrication, on the other hand, the contact geometry changes because the surfaces deform under the high pressure. Thus, the measurement must be able to deal with contact geometry changes.

The film thickness results will be useful for ranking lubricants. Absolute values will not be relevant since the exact application contact conditions can usually not be fulfilled in the testing environment. Still, the measurements should be carried out at contact conditions, for example contact pressure and sliding velocity, as close to the application conditions as possible.

6.3.1 Electrical Methods

One way to verify if there is a lubricant film that completely separates the surfaces is to use a simple electric circuit (see Figure 6.3). If the metal surfaces are in contact the electric circuit is closed and a current can be recorded. If there is an electrically insulating lubricant film between the surfaces, the circuit is broken and no current is recorded. This *conductivity measurement method* is the basis for different condition monitoring systems, where the main

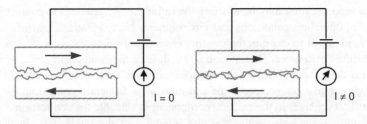

Figure 6.3 Electrical circuit set-up for monitoring the presence of metal–metal contact

focus is to detect any metal–metal contact that can indicate initialized damage, for example between the roller element and the race of a bearing.

The next step is to measure the thickness of the lubricant film, for example the separation gap between two solid bodies. This can be done with the use of sensor systems that combines *inductive* and *capacitive sensors* with conductivity measurements. Both inductive and capacitive sensors are noncontacting sensors. They are mounted on one of two mating surfaces and measure the distance to the other. This means that the lubricant film thickness is measured at one specific point or location of a wider contact area, for example a point within a hydrodynamic lubricated journal bearing.

The capacitive sensor is preferable when the gap is almost constant, since it can register very rapid changes. The inductive sensor is most efficient where the gap changes continuously, since the sensor registers changes of metal (volume or mass) within the induction field. When using these electrical techniques, it is of course also important to know the electrical characteristics of the fluid, and very careful calibrations are needed.

6.3.2 Optical Interferometry Method

If oil is poured on top of water, the oil spreads out, forming a thin oil film (see Figure 6.4). Rainbow colours are reflected when light shines on the water. If the oil film is very thin, bright and clear colours appear. By knowing the wavelength of each visual colour, the thickness of the oil film can be estimated closely. The basis for the optical interferometry film thickness measurements method is analogous to the oil on water description.

The light illuminating the surface can be reflected from the surface of the glass or from the surface of the lubricant (see Figure 6.4). The light that is reflected by the glass surface will

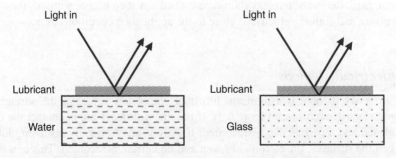

Figure 6.4 Illustration describing the optical interferometry principle

travel a longer distance than the light reflected by the lubricant surface. By interferometry the difference in travelling distance and thus also the thickness of the lubricant film can be estimated from the colours that are formed. This method can be used to measure only thin lubricant film thicknesses.

The thickness of the lubricant film between a rotating disc and a freely rotating steel ball can be measured by optical interferometry in the test set-up shown in Figure 6.5. The rotating disc is made of glass or in some cases sapphire. To enhance the reflection of the incoming light, a layer of chromium covers the disc. A steel ball, supported by three rollers, is freely rotating

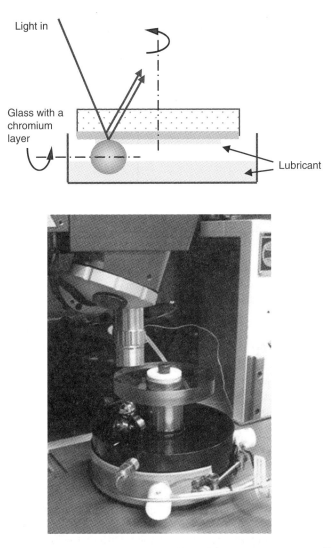

Figure 6.5 Ball-on-disc set-up for interferometry film thickness measurements. Illustration of the method principle and a close-up photograph of the test rig

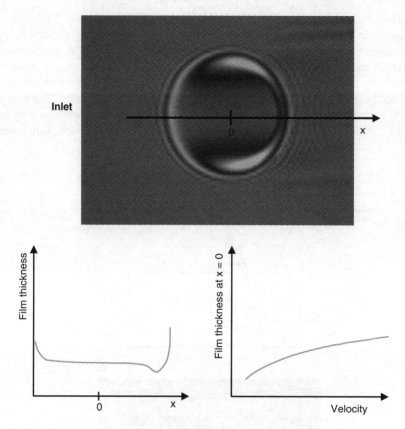

Figure 6.6 Interferograph (interferometry photograph), calculated lubricant film thickness from the interferograph (bottom left) and film thickness as a function of velocity (bottom right) calculated from several interferographs (For a colour version of this figure, see the colour plate section)

against the disc. When the disc rotates the steel ball also rotates. The lubricant is pulled up due to the movement of the steel ball, forming a thin lubricating film underneath the glass disc.

The film thickness measurements can be shown as pictures (see Figure 6.6). The reflected colours indicate different film thicknesses over the contact area. The rotating disc moves from the left to the right in the picture and the photo is taken from above through the glass disc. The lubricating film thickness is always thicker at the inlet and thinner at the outlet. The graph shows the film thickness along the central horizontal line indicated in the picture. Generally, the central film thickness ($x = 0$) is selected to represent the film thickness.

The applied load by the steel ball against the glass disc can be varied. There is a risk of breaking the glass or scratching the chromium layer if the applied load is too high. Therefore, the method allows measurements to be made with limited applied loads and consequently limited contact pressure ranges.

The velocity of the rotating disc can also be varied. Different velocities will give different film formation behaviours. Plotting the central film thicknesses versus the velocity will give information in a broader working range than one single image of the whole contact.

6.4 Tribological Evaluation in Mixed and Boundary Lubrication

The friction and wear behaviour reveals information of the contact performance during mixed or boundary lubrication. Friction and wear can be evaluated in a number of tribological test methods. The choice of test depends primarily on the requested operating conditions and contact geometry. The environmental conditions may be controlled if the test set-up is, for example, equipped with a hood that allows the humidity and the temperature to be controlled. How to plan and select a model test is described in Section 6.2. In this section a number of common model test set-ups are presented.

The friction force is generally measured during the test. In most cases the friction force can be monitored all through the test. The level of friction force may vary during a test, for example it may increase or decrease during running-in and then reach a stable level. Friction results may be presented in different ways, for example as the coefficient of friction at the stable level or at the end of the test, or as a graph showing the development with time, sliding distance or number of cycles.

Designing a set of experiments requires decisions on which appropriate conditions should be chosen. It often ends up with a compromise between total time for the whole experiment and the optimal length of each test run.

Wear can be measured or calculated from the weight loss, the worn-off volume or the change in surface roughness. The weight loss or worn-off volume are commonly related to the sliding distance, number of cycles or passages of contact, or test time. Wear can also be evaluated visually or microscopically, for example by scanning electron microscopy (SEM), revealing the wear and friction mechanisms. (Methods for microscopy and chemical analysis will be covered in detail in Chapter 8.)

6.4.1 The Pin-on-Disc Tribotest

The pin-on-disc test is probably the most commonly used tribotest. It is simple and easy to use for friction and wear rate evaluation of sliding contacts. The method is versatile, allowing both dry and lubricated contacts to be examined. This, in combination with the relatively low price of testing, often makes it the number one selection among tribotests.

The test materials are chosen from the application of interest. Normally, the material that is expected to be more wear resistant is chosen for the disc. If not, manufacturing limits implies other choices. If the less wear-resistant material is chosen for the pin, there will be more wear, which gives more accurate measurements of the amount of wear for a given test.

The pin-on-disc has an upper test pin, which is loaded against a rotating test disc. The sliding motion is unidirectional and the sliding velocity is varied when the rotating velocity or the track radius is altered. The applied load can also be varied. The pin and disc can be immersed in the test lubricant. An illustration of the set-up and a close-up photo are shown in Figure 6.7.

The geometry of the upper test specimen can be changed to run the test under different contact conditions. A higher contact pressure is obtained by changing the pin with a flat (distributed) contact area to a ball with a point contact area. During testing with a ball, the point contact area will be transformed into a distributed contact area when it wears down. It is important to note that only small geometries can be tested in the pin-on-disc test set-up due

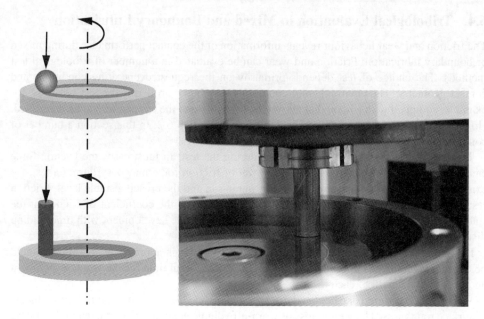

Figure 6.7 Illustration of ball-on-disc (top, left) and pin-on-disc (bottom, left) set-ups. The upper specimen is held fixed while the lower disc rotates. Close-up view of the pin-on-disc apparatus. The pin and disc can be immersed in a test lubricant

to the dependence of velocity on the radius, that is large geometries giving large contact areas experience uneven wear behaviour.

The friction force versus time is the output from the test. The friction force is continuously measured during the test by using a strain gauge force transducer mounted in the pin holder.

The rotational speed can be changed during a test, yielding, for example, a continuously increasing sliding velocity, where the coefficient of friction versus sliding velocity can be measured (see Figure 6.8).

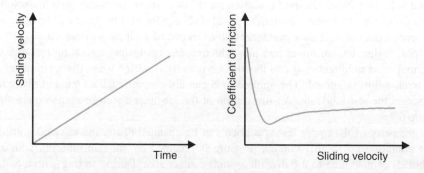

Figure 6.8 Example of recordings from a pin-on-disc tribotest. Sliding velocity versus time (left) and the corresponding friction coefficient versus sliding velocity (right)

Figure 6.9 Illustrations of the wear volumes for a cylindrical pin and a pin with a spherical tip respectively

The pin (or ball) is in contact with the disc all the time, while different points of contact appear for the disc following the circular track. The pin will therefore be continuously worn, while the circular track will be worn when the pin is passing over that part of the track. In some test set-ups a linear potentiometer measures the movement of the pin piston and can provide in situ wear measurements, with a resolution of about 1–2 µm. The wear of a pin with uniform cross-section and a pin with a spherical tip is schematically shown in Figure 6.9. The wear of the rotating disc can also be measured, although this can be more difficult than for the pin. Worn-off material may accumulate in the lubricant. These particles may be brought into contact, thereby disturbing the test by, for example, abrasive wear.

6.4.2 The Reciprocating Tribotest

In the reciprocating tribotest, the moving upper specimen is loaded against a lower stationary specimen with the use of, for example, a spring system (see Figure 6.10). The reciprocating motion can be provided from a mechanical system or by linear electrical motors. As for the pin-on-disc set-up only the upper specimen is in continuous contact.

Different geometries can be selected for the upper specimen, which may yield distributed, line or point contact areas. Specimens may be produced from real components or small real components may be used, for example parts of a piston may be tested against a cylinder lining. Another advantage of the reciprocating tribotest is that very large contact areas can be tested. The reason for this is that the reciprocating movement is linear and the sliding velocity does not differ within the contact area.

The applied load, the sliding velocity and the stroke length may be varied between tests. Very small strokes may be used to simulate fretting. The tests may be run dry or lubricated. The specimens are surrounded by the test lubricant, which can be temperature controlled. Generated wear particles may be pushed out of the contact and occasionally be brought back into the contact during a test. If the worn-off particles remain in the lubricant and are brought back into the contact this may give rise to additional abrasive wear.

The wear rate for reciprocating wear tests is normally evaluated from the size of the wear scar of the moving specimen, for example using a profilometer (see also Section 8.3).

Figure 6.10 Illustration of reciprocating test set-up, where lower specimen is stationary and upper specimen is reciprocating. Reciprocating tribometer with lubricant

The sliding velocity is zero at the turning points for reciprocating motions. Different types of motion are possible. It is usual to apply a sinusoidal velocity, where the velocity increases and reaches a maximum at the centre of the stroke and then decreases again (see Figure 6.11).

The friction force is also zero at the turning points and switches between being positive and negative depending on the direction of motion (see Figure 6.12). If the static (i.e. at zero velocity) and dynamic friction coefficient (i.e. as soon as there is a relative motion between the surfaces) are identical and the dynamic friction is independent of the sliding velocity, then a quadrant shape of the friction force versus time is expected. The friction force is used to calculate the coefficient of friction and the centre value is commonly chosen to represent each stroke. These values of the coefficient of friction can be plotted as a function of time.

6.4.3 The Twin Disc Tribotest

In the twin disc tribotest two identically sized discs rotate against each other. The discs are in line contact under unidirectional motion. The disc materials can be varied, for example different types of steel can be tested, but generally testing is performed using identical materials. The test set-up is shown in Figure 6.13 together with a close-up photo. The lubricant is dropped on to the twin discs and pushed into contact by the rotation. The lubricant can either be disposed of or recycled after the contact. There is a risk of abrasive wear and/or fatigue if the

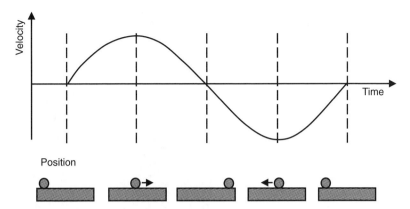

Figure 6.11 One working cycle for the reciprocating test. The velocity as a function of time is shown in the graph. The corresponding position of the upper specimen at different stages of the working cycle are indicated

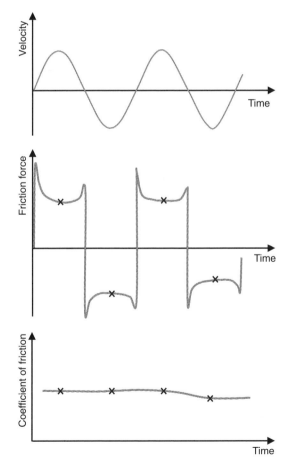

Figure 6.12 Typical results from a reciprocating test. Velocity versus time, that is the input signal (top). Continuously measured friction force versus time, where sampled values are shown (middle). Calculated coefficient of friction values versus time (bottom). In this example two friction force values of each working cycle, one in each sliding direction, are used

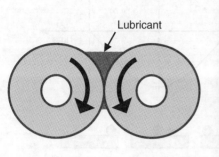

Figure 6.13 Illustration of twin disc set-up. Close-up view of test specimen in the twin disc set-up. Lubricant may be dropped on to the discs (For a colour version of this figure, see the colour plate section)

lubricant is recycled without filtering. The twin disc can be used to, for example, simulate gear teeth contact.

The twin discs can be rotated independently of each other, allowing pure rolling, pure sliding or a combination of both. The standard procedure of the twin disc set-up states that the discs should rotate in opposite directions. During pure rolling conditions the same point of contact on each disc will always meet since the discs have identical sizes and the rotating speed is equal and constant. A combination of rolling and sliding may occur when the speed and/or sizes of the discs differ. Pure sliding also occurs by having the discs rotate in the same direction, where one of the discs is allowed to rotate at a velocity close to zero. This is an exception to the standard test procedure, which yields high wear rates. The applied load and the rotational speed of the two discs may be varied. The applied load is measured with a force sensor on the load lever. The friction force is calculated by continuously measuring the traction force between the disc specimens using a torque sensor. The coefficient of friction is calculated, knowing the friction force and the applied load (refer to Chapter 1). Typical results from the twin disc test, that is coefficient of friction versus time, are shown in Figure 6.14.

Pure rolling   Slip   Pure sliding

Figure 6.14 Rolling and sliding visualized for two discs in contact. Pure rolling is shown to the left and pure sliding to the right. In the middle a situation with combined rolling and sliding is shown

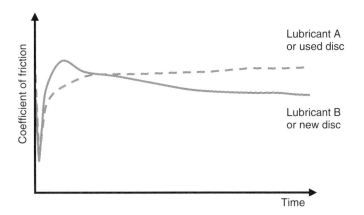

Figure 6.15 Different running-in behaviour is visualized by using two lubricants with new and used discs correspondingly

Tests are usually run at constant rotational speeds. Different lubricants may show different frictional behaviour (see lubricants A and B in Figure 6.15). The coefficient of friction may initially be high and then it reaches a lower, stable level. This behaviour is typical for running-in of the discs, as shown by the continuous curve in Figure 6.15. For discs that have already been run in, typically the coefficient of friction is low in the beginning of a test and then increases with time, as shown by the dashed curve in Figure 6.15.

The wear rate of the discs is generally low. Wear reduces the radius of the discs. In some test set-ups this radial change can be measured in situ (see Figure 6.16). However, if the wear rate is low the radial movement is correspondingly low, leading in most cases to unreliable in situ measurements of the wear. However, in some tests the wear rate may increase with time, leading to catastrophic wear and failure. This is shown as a quick radial reduction of the discs. If possible, calculating the wear rate from weight loss is both very reliable and common.

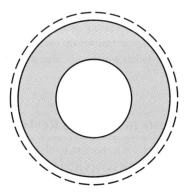

Figure 6.16 Radial wear of a twin disc is shown where the dashed line indicates the original surface

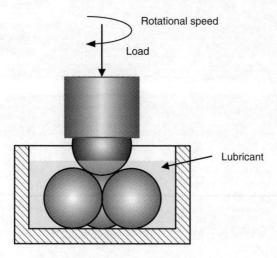

Figure 6.17 The four-ball rotary test set-up

6.4.4 The Rotary Tribotest

The rotary tribotest is a versatile test for evaluating seizure and wear. In this section we will cover the four-ball set-up only.[1] The upper holder has one rotating steel ball, which is loaded against three stationary lower steel balls. All contact areas are drowned in the test lubricant. The load can be applied by using deadweight or through a hydraulic system. The rotation is central along the symmetry axis of both the upper and the lower holders. The principle for a four-ball set-up is presented in Figure 6.17.

The material used in the four balls is commonly bearing steel since the focus mostly is to evaluate the performance of lubricants. Thus, material investigation is of minor importance.

Friction can be monitored in the rotary tribotest, but is seldom used. Instead the critical loads that lead to initial seizure and welding are of interest. The initial seizure load is defined as the load where a significant wear scar is observed. The maximum load is achieved when the upper steel ball is welded against the lower three steel balls. During a test the load is varied from low load to very high load under unidirectional relative motion. After each load step, the surfaces undergo visual inspection and the initial seizure or the weld load can be identified.

Wear can also be tested using the same test set-up under moderate applied loads. Circular wear scars will appear on the lower balls, while a circular wear track will appear on the upper ball (see Figure 6.18).

6.5 Selection of Model Tests to Simulate Real Contacts

When testing materials and lubricants it is important to plan critically how to perform the tests. In some cases a field test, a bench test or a component test may be required. In this section we will focus on the criteria for selecting a model test.

---

[1] Standard test procedure: ASTM D4172.

Figure 6.18 The wear is measured from the rotary tribotest. The wear mark of one of the lower balls is shown to the left and the wear scar of the upper ball is visualized to the right

Three applications will be considered to aid the reader through the thinking for choosing relevant model tests. The applications are *hydraulics*, *gears* and *combustion engines*. All three applications operate in either the mixed or the boundary lubrication regimes and have been covered in more detail in Chapter 5.

6.5.1 Hydraulics

In a hydraulic system the main issue for the fluid is to transmit energy. At the same time the fluid supports lubrication in all parts of the system. A simple hydraulic system consists of a hydraulic pump, hoses, valves and an actuator, that is the hydraulic motor or cylinder. The parts that need lubrication are the moving parts in the pump, valves and motor or cylinder.

Two different lubricated parts of the *axial hydraulic pump* have been selected to exemplify model testing. These are the piston–ball joint at the rear end of the piston and the piston ring–cylinder liner contact at the front end of the piston, as shown in Figure 6.19. The lubrication requirements of these parts differs. In the piston–ball joint there are high loads and relatively small movements. In the piston ring contact there are moderate loads and higher relative velocities.

The piston–ball joint contact is a conformal contact with mainly sliding motion, making the *pin-on-disc tribotest* run in the mixed lubrication regime a suitable test method.

The piston ring–cylinder lining is a sliding contact operating in the mixed lubrication regime at low to moderate temperatures. It is interesting to study the wear rate of the piston rings, which is suitably done in a *reciprocating tribotest*.

6.5.2 Gears

Gears operate in the mixed and the boundary lubrication regimes with a combination of rolling and sliding motions. Lubricant formulation is done to handle both regimes with a strong focus on the boundary lubrication regime.

The FZG[2] is a bench test for evaluating both gear performance and gear lubricants. The FZG has carefully designed gears with well-defined surfaces. During testing the applied load is gradually increased and the wear scars are evaluated after each load stage. Since the FZG is expensive and time-consuming to run, it is sometimes more suitable to run a simpler model test.

---

[2] Forschungsstelle für Zahnräder und Getriebebau.

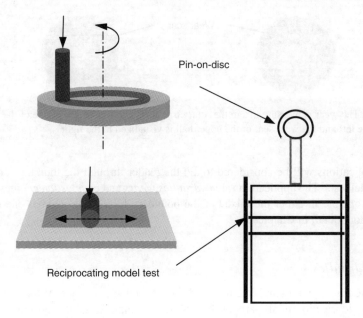

Figure 6.19 The lubricated contacts in a hydraulic pump can be modelled with different tests. The piston cylinder contact can be modelled by a reciprocating model test and the piston ball joint contact can be modelled with a pin-on-disc test

For simpler tests the twin disc and the rotary tribometer test methods can be used (see Figure 6.20). The rolling and sliding in a gear contact can be simulated with a *twin disc* set-up and wear can be evaluated. Discs of different materials and sizes can be used.

Gear lubricants are formulated with extreme pressure additives to avoid welding of the gears. Therefore, weld loads are usually tested. This can be done using the *rotary tribometer*, where high enough contact pressures can be applied for welding to occur.

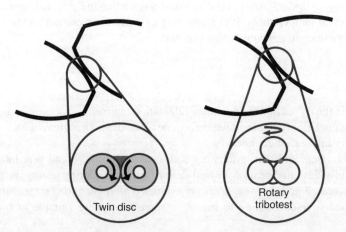

Figure 6.20 The rolling and sliding performance in a gear contact can be modelled with a twin disc test (left) and the weld load with a rotary tribotest (right)

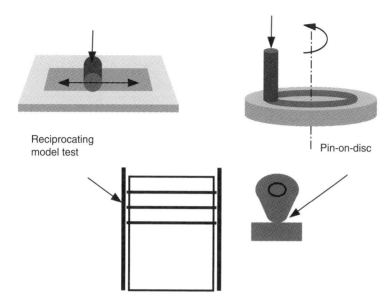

Figure 6.21 The piston–cylinder contact can be modelled by a reciprocating test (to the left) and the cam–follower contact can be modelled with a pin-on-disc test (to the right)

6.5.3 Combustion Engines

The combustion engine is lubricated with one lubricant operating in the boundary to the full-film regime. Two parts of a combustion engine have been selected to show the use of model testing (see Figure 6.21). The piston–cylinder liner and the cam–follower represent two different types of lubricated contacts in an engine. Due to the combustion in the engine, the piston–cylinder is exposed to very high temperatures. The cam–follower is exposed to much lower temperatures.

The piston–cylinder liner shows a start–stop phenomenon at the turning points of the piston, that is reciprocating motion. The lubrication regime changes from boundary to mixed lubrication for each stroke. Therefore, the *reciprocating tribotest* is chosen. The environment can be controlled and monitored by adding a hood to the tribotest. This will make it possible to run tests at temperatures relevant for the application.

The cam–follower is a sliding contact with different velocities in both mixed and boundary lubrication. This contact is usually difficult to lubricate due to uneven wear rates of the cam. Therefore, it is very interesting and important to find materials and lubricants that keep the friction and the wear rates at a minimum. This can be simulated in a *pin-on-disc*, where a flat surface contact is in contact under unidirectional relative motion. Both materials and lubricants may easily be evaluated under different conditions.

6.6 Summary of Tribotest Methods

The possible types of contacts and motions that can be used in the tribotests presented in this chapter are summarized in Table 6.1.

Table 6.1 Summary of possible types of contact areas and motion in the presented tribotests

	Pin-on-disc tribotest	Reciprocating tribotest	Twin disc tribotest	Rotary tribotest
Contact area				
Point	x	x		x
Line		x	x	
Distributed	x	x		
Motion				
Sliding	x	x	x	x
Combination			x	
Rolling			x	
Direction				
Unidirectional	x		x	x
Reciprocating		x		
Main benefits	+ Easy to use + inexpensive + dry and lubricated test + easy to change material in both upper and lower specimens + 'world standard'	+ Alter contact geometries + dry and lubricated test + easy to change material in both upper and lower specimens	+ Symmetrical geometry + dry and lubricated test + easy to use + identical material in both specimens	+ Standard test set-up for lubricant tests

References

[1] Czichos, H. (1992) Design of friction and wear experiments, in *Friction, Lubrication, and Wear Technology*, vol. 18, ASM Handbook (ed. P.J. Blau), ASM International.
[2] DIN Standard 50322 (1984) Wear – Specification of the Categories of Wear Testing, DIN.
[3] Hogmark, S. and Jacobson, S. (1991) Hints and guidelines for tribotesting and evaluation. *STLE Lubrication Engineering*, **48**, 401–409.

7
Lubricant Characterization

Lubricants are expected to hold the quality indicated on the container. Focusing on the quality during development and production ensures this. The quality can be kept by having a good process for lubricant characterization all the way through development and production, and to the application where the customer will use the product.

Lubricant characterization is more than just analysing a sample. It requires some general characterization concepts to be covered before the actual analyses of nonused and used lubricants can be explored.

7.1 General Characterization Concepts

Characterization can be done both qualitatively and quantitatively. In other words the characterization can describe *what* and *how much* is in the lubricant. Describing what is in the lubricant can be done to different extents. A complete qualitative analysis would respond to the exact types of hydrocarbons and other chemistries present in the lubricant. A complete quantitative analysis would give the percentage of, for example, carbon, hydrogen and oxygen. A partial analysis would give the type or amount of selected constituents (e.g. phenolic antioxidants) in the sample. Characterization also involves analysing both physical and chemical properties (described in Chapter 2). The latter could include both chemical composition and chemical structure.

All characterization requires work to be done ahead of the actual analyses being run. This work includes the following steps: planning, sampling, analysing the sample, doing calculations and evaluating the results. In some cases diluting the sample may be relevant to allow for more analyses to be made from a small sample [1].

7.1.1 Planning

Planning is an important step in the analysis work. It involves understanding the requirements in the development process, the environment in production and the operating conditions for a lubricant being used. Prior to analysing it is important to understand:

Lubricants: Introduction to Properties and Performance, First Edition.
Marika Torbacke, Åsa Kassman Rudolphi and Elisabet Kassfeldt.
© 2014 John Wiley & Sons, Ltd. Published 2014 by John Wiley & Sons, Ltd.

- The information needed.
- How to retrieve the information needed and to which accuracy.
- The method to use in order to retrieve the information needed to the desired accuracy.

The information needed can sometimes be difficult to pinpoint. Here a dialogue between the laboratory performing the analyses and the receiver of the output is important.

When the information needed has been identified it is important to understand fully which accuracy is necessary. This may be crucial both from a cost and a time perspective. If too accurate results are requested, the analyses may take a very long time to do and cost more than necessary. On the other hand, if the accuracy is too low the information retrieved may not be relevant. Therefore, it is important to understand the accuracy need.

Selecting a relevant method to give the information needed to the agreed-upon accuracy is important, since there are usually several methods available. Selecting the most beneficial method for analysis is the responsibility of the laboratory.

7.1.2 Basic Mixing Theory

To understand sampling, it will be useful to cover some basic mixing theory, exemplified by blending in a stirred tank, using a propeller creating a flow in the propeller axis direction (see Figure 7.1). The propeller pushes the fluid downwards in the tank creating a semi-circular flow in the stirred tank, where the fluid is moved from the bottom of the tank to the side, and back to the propeller region from above. All fluids (including base fluids, high viscosity additives and liquid additives) and solids (e.g. solid additives) in the tank follow the same flow pattern, even though the radii of the semi-circular flow may vary. The highest mixing intensity is in the region of the propeller where the flow is pushed to the bottom. The lowest mixing intensity is at the liquid surface where the velocity is zero. Mixing, time and heat are required to blend the base fluids with the additives [2]. An increased temperature will in most cases reduce the

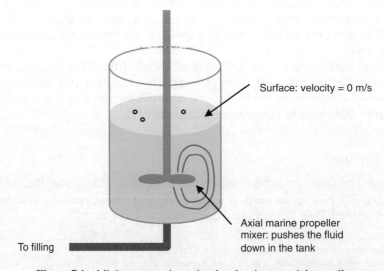

Figure 7.1 Mixing pattern in a stirred tank using an axial propeller

blending time, since the viscosity decreases with increased temperature and lower viscosity improves the mixing. Also, some solid additives require heat to dissolve.

The solid additives are soluble in the base fluids. Due to varying additive chemistry the dissolving can be anything from fast to slow. Also, the way solid additives are added may affect the time of dissolution. If solid additives having low density are added on to the top surface of the base fluids, mixing will be poor. Instead, such additives should preferably be added to the bulk of the stirred tank. Solid additives with high density may remain on the bottom of the stirred tank if the tank geometry is not optimized from a mixing perspective or if mixing itself is insufficient.

Lubricants consist of base fluids with blended or dissolved additives. A lubricant should be stable and not give rise to separation of any kind, such as additives crystallizing. However, mixing of lubricants in an application may still involve different phases due to water contamination or particles in the lubricant. Thus, mixing phenomena in lubricants may occur and be relevant also during use.

7.1.3 Sampling

In most cases a small test sample and not the full volume is collected and brought to the laboratory, that is the sample is always smaller than the total volume. When taking a sample it is important to consider the volume necessary to perform all the analyses planned and required to retrieve the relevant information. The sample volume can be decided upon in a discussion with the laboratory carrying out the analyses. Taking two samples will increase the probability of receiving more accurate results and also will allow for statistics to be calculated.

Since a sample is part of the total volume, it is relevant to understand the importance of collecting a sample that is representative of the full volume. If the sample is not representative, the measurements will not be relevant for the blend or the product. This is crucial for the remaining analysis work. Despite this it is too often neglected or not fully understood.

Samples taken should be representative of the large volume and be large enough for their purpose. Sampling from a stirred tank, for example in production, requires understanding of the mixing pattern and the time for blending (see Figure 7.2). Samples can be taken from different positions in a stirred tank. However, due to practical purposes the number of positions may be limited. For instance, taking a sample from the surface may give faulty results due to the surface velocity being zero, causing reduced mixing. Therefore, a sample taken from the surface may not be fully blended, giving rise to a nonrepresentative sample.

Taking a sample from the bulk closer to the marine propeller will give a more accurate sample since the mixing is more intense in this area. This will ensure that good mixing has taken place. However, high viscosity additives may still be streaky and solid additives may still be undissolved if the blending time has been too short. Thus, sampling needs to be done after mixing has been completed. However, waiting too long with sampling increases blending time and thereby production costs.

Sampling for used lubricants requires an understanding of the application itself. In general, a sample should be taken when the lubricant is in use to avoid, for instance, settlement of particles and sediments. Also, the sample should be taken preferably where there is a flow of the lubricant. Avoid taking a sample from free top surfaces since there may be foam, dust or other contaminants there.

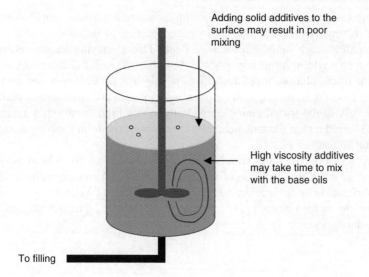

Figure 7.2 Visualization of areas where sampling may be difficult (For a colour version of this figure, see the colour plate section)

Equally important to taking a representative sample is having clean sample containers. If the sample container is contaminated the analyses will most likely give faulty results. For example, it is crucial to have clean sample containers if particle counting should be performed.

When the sample is taken it is good practice to label the container with the product name and date. In the case of development samples, the product identifier may be the blend number. For production purposes the product name can be supplemented with the batch number to keep a good audit trail. It might be useful to keep used oil samples marked with product and customer names. Samples can be saved if it is believed that they may serve a later purpose. The time for keeping samples may vary between sample types.

7.1.4 Diluting the Sample

It is often convenient to take a small sample from an application and send it to a laboratory. Sometimes the user of the application would like to have many analyses run even though the sample volume generally would not allow it. In those cases diluting the sample, that is increasing the volume by blending the sample with, for example, a solvent or a base fluid, may allow for more analyses to be run, giving more information about the sample.

It should be noted that every time the sample is diluted there is a risk of losing information due to contamination or improper blending. This will give rise to less accurate measurements and a risk of making inadequate decisions about the lubricant quality. Therefore, it could be more relevant to be more precise regarding the information need and the accuracy required. The number of analyses could in that case be reduced where the sample volume might be large enough. Also, the order of the analyses could be relevant where some analyses can be run using the same volume of the sample without losing information.

7.1.5 Collecting Analysis Data

There are several definitions to cover when discussing measurements. Among these are accuracy, precision, reproducibility and repeatability. In addition, it is relevant to cover the terms absolute values, analysis values in relation to a reference value as well as trend analyses.

A *reference value* is the true value of the product. The reference value is the value measured on the final formulation used for the product. The *accuracy* involves the accuracy of the measurement guaranteeing calibration of the instruments used and also a measured deviation due to possible changes in the sample (e.g. contamination, ageing). Thus, the shift between the reference value and the measured value depends on the real difference and the accuracy of the measurement. In Figure 7.3, the value of the measurement is shown on the y axis and the reference and measured values on the x axis. When multiple analyses are done on one sample the measurements will naturally spread. Then a discrete value may represent a mean value. The *precision* in the measurement is the total spread of the results. The closer the measurement results, the better is the precision of the analysis itself.

The precision also depends on the circumstances for the measurement. If one person measures a sample value using the same instrument in the same laboratory, the precision is referred to as *repeatability*. If the measurement is performed regardless of circumstances, that is regardless of the person carrying out the measurement, the laboratory and so on, the precision is referred to as *reproducibility*. If a sample is sent to different laboratories doing the same analyses, the result will give the reproducibility. However, if the same person at the same laboratory reruns the sample, repeatability will be shown.

It is common that identical samples are sent to several laboratories in order to ensure that measurements in defined instruments can be relied upon regardless of the laboratory used. This ensures that calibration of instruments using standard samples with defined contents is done at the laboratory. Laboratories can be certified, which will guarantee that the instruments are calibrated regularly and analyses are performed in the correct way.

Measurements may be carried out in different ways. An *absolute value* can be measured and reported where the sample is measured in an instrument, which has been thoroughly calibrated. The absolute value may be measured to, for example, 1000 ppm of phosphorus

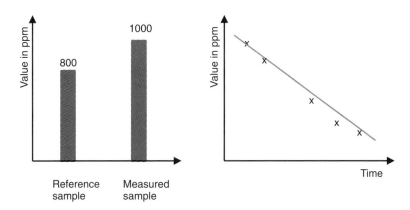

Figure 7.3 Different ways of presenting measurements: a measured value in relation to a reference value (to the left) and a trend analysis (to the right)

(see Figure 7.3). The same sample can be run and *compared with a reference sample* with the known, correct composition for the same lubricant. In this example the reference value of phosphorus shows 800 ppm whereas the measured sample shows 1000 ppm. Thus, the measured sample has 125% of phosphorus and can be described as such compared to the true reference sample.

If samples are taken at different times it will be possible to show a *trend of values*. This can be useful when describing, for example, the remaining life of a lubricant. In the case below the phosphorus content decreases with time and has consequently been consumed.

7.1.6 Calculations and Evaluation

Taking one sample and analysing it once will give one discrete measurement value. Several samples and several analyses will give a range of values, allowing for the mean value and the corresponding standard deviation to be calculated.

Calculation of results will be an important final action of the laboratory that together with the evaluation of the results will give valuable information. The trained analyst will be able to add valuable information to the measurements, indicating where to be careful and areas to consider for future use and future analyses.

7.2 Condition Analyses of Lubricants

An initial fast or manual inspection of the lubricant sample will provide a lot of information for an experienced eye and nose (see Figure 7.4). If visual inspection is possible, it will give valuable first-hand information. Light coloured lubricants are easily inspected while dark brown or black coloured ones are difficult or in some cases nearly impossible to inspect visually.

Many lubricants are bright and clear when they are new, indicating good condition. If this is not the case the trained eye will look for sludge, particles and/or haziness. This is relevant for both nonused and used lubricants. For nonused lubricants this will imply development or production challenges and for used lubricants there could be several reasons that may need a deeper investigation. For example, sludge may imply oxidization, particles may mean contamination and haziness may show water entrainment.

The smell of the sample may indicate contamination and shaking the sample will give information about, for example, dilution or thickening.

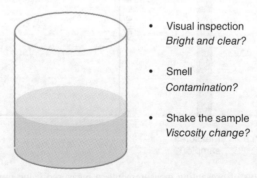

Figure 7.4 The first steps in analysing the lubricant sample

Lubricant Characterization

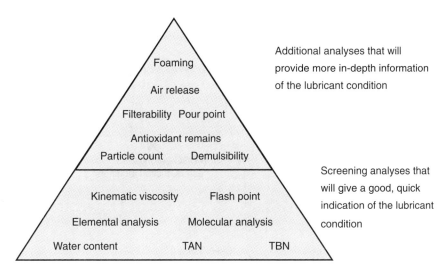

Figure 7.5 Examples of lubricant analyses (For a colour version of this figure, see the colour plate section)

The initial fast inspection will be supplemented by analyses to get both quantitative and qualitative information. Different analyses are utilized depending on what information is required. The most common analyses can be summarized, regardless of lubricant type and whether nonused or used.

In order to get a broad picture one usually analyses the kinematic viscosity, flash point, total acid number (TAN), total base number (TBN), elemental and molecular composition, and water content (see Figure 7.5). These analyses are relatively quick to run and will give a good indication of the lubricant condition. If considered necessary, these basic analyses will be supplemented with more in-depth and thorough ones, such as pour point, foaming, filterability, air release, demulsibility, particle count and remaining antioxidant analyses. They will give more detailed information about the condition and will be used to pinpoint areas of problems. The analyses will enable recommendations on, for example, changing the lubricants or cleaning the system in order to prolong the life of the lubricant and the application. How to analyse the different chemical and physical properties of lubricants has been covered in Chapter 2 and tabulated in Table 2.1.

Elemental analyses can be carried out by different methods [3]. The most common ones are the X-ray fluorescence spectroscopy (XRF) and inductively coupled plasma (ICP) methods. In XRF the lubricant is irradiated by X-ray photons. The interaction between the incoming photons and the atoms results in emission of characteristic X-ray photons, that is X-ray fluorescence. The emitted X-ray photons are detected and quantified. This method is useful for detecting, for example, barium, calcium, phosphorus, sulfur and zinc, which are common additive elements. With the ICP method the lubricant is diluted in the solvent and injected into a plasma, where it becomes atomized. The elements in the lubricant are detected, for example, from optical emission from the atoms in the plasma and quantified. This method is useful for detecting a broader spectrum of elements at lower concentration levels, such as iron, lead, tin, antimony, molybdenum and boron.

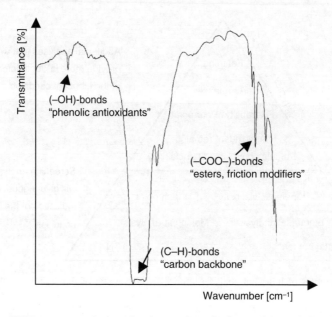

Figure 7.6 An FTIR spectrum of a combustion engine oil. Some characteristic absorption peaks, corresponding to characteristic molecular bonds, are indicated

Molecular analyses, like Fourier transform infrared (FTIR[1]) spectroscopy, are also used to evaluate lubricants. With FTIR spectroscopy, IR light of different wavelengths irradiates the lubricant. IR light activates molecular vibrations. Absorption of particular wavelengths indicates the presence of characteristic molecular bonds. Consequently, different functional groups such as esters, alcohols and fatty acids can be qualitatively and quantitatively determined (see Figure 7.6).

Evaluation of the condition will be simplified if the current properties can be compared with the reference properties. This is relevant for all the suggested analyses above.

7.3 Nonused Oil Characterization

Nonused oils are characterized both during development and in production. The procedures and differences will be described in this section. Lubricant formulation, and indirectly the development process, has been covered in Chapter 5.

7.3.1 Development

The development plan of a new product includes the analyses to be run in order to fulfil the requirements of the application. This implies that planning the analyses to be run is part of the development process. The development of a new product is an iterative process based on knowledge of the application requirements and operating conditions as well as a knowledge of base fluids and additives and the interaction between different additives and between additives, base fluids and surfaces.

---

[1] The FTIR method will be further described in Chapter 8.

Development is commonly done in small scale. Formulations consist of base fluids and additives. Some base fluids or additives may have a very high viscosity and some additives may be solid. Regardless of the type of additive, blending is still simple in small scale since heating is quick and mixing times are relatively short. Also, the degree of blending can easily be visualized in see-through beakers.

Sampling is normally easy to handle during development, since small volumes are often used and the total volume will serve as the sample. During development it is crucial to have clean beakers for the lubricant since the analyses made will finally serve as the reference of the product being developed.

The quality limits for the product need to be defined during development. These limits will set the allowed tolerance for the product.

Developing lubricants for hydraulics, gears and combustion engines has been covered in Chapter 5 and will therefore not be covered here. Analyses run during development are according to the requirements.

7.3.2 Production

Successful mixing of base fluids and additives during production requires more blending competence and experience than during the development phase. The production scale may require thorough blending instructions to consider the scaling-up of the blending, which will form the basis for the planning of the analyses to be performed. This should allow for the best way of mixing the base fluids with the additives, thus ensuring the final blend to be within the set quality limits.

It is crucial to ensure that the correct base fluids and additives have been used. For the product to fulfil the requirements for the application, all additives, including solid and highly viscous ones, must be completely mixed with the base fluids. Here sampling position and timing are important. If the sample is taken at a faulty position or too early, it may also cause an extra sample to be needed. Analyses will also reveal contamination of the product. Contamination can be from a previous product in the blender or by faulty base fluids and additives.

The raw materials, that is base fluids and additives, are analysed on receipt at the production unit. If this is done, any faulty results from the blending will not depend on raw materials not being according to the specification. Thus, this source of error can be taken away during the investigation if blending is unsuccessful.

The production sample is inspected using both the eyes and the nose. This will give first-hand important information about the quality. After that, the product blended will be within the set quality limits for the product. The tolerances for the characteristics in production are based on knowledge about the application. Each product is released according to an accepted tolerance. To accomplish this, a plan for the required analyses to be run is needed and used to ensure product quality. Production-related techniques apart from basic mixing and sampling will not be covered in this book even if it may be relevant to the product quality.

The purpose of analysing prior to filling is to ensure that the correct product is filled in the right container. It is equally important to ensure that the product being filled is not contaminated. Filling the right volume can be done by weighing, using the actual density of the product.

Sampling and testing for the filling analyses are usually fairly simple. A sample is taken from the filling tube and very few analyses are needed to ensure that the correct product is being filled and that it is not contaminated. Commonly viscosity and molecular analyses are run where viscosity will secure the anticipated viscosity of the product and FTIR spectroscopy will catch any contamination prior to filling.

7.3.3 Application Examples

Indoor *hydraulic oils* contain 99% base fluids. In order to function properly in the application they must not be contaminated. Contamination may give rise to reduced air release values and possibly cavitation, foaming or reduced water separation. Thus, producing hydraulic oils requires extremely clean blenders in order to avoid or at least reduce contamination.

Having good procedures in place for production will simplify the production of hydraulic oils, ensuring high quality products.

Typically, viscosity and density are measured and elemental and molecular analyses are run. The viscosity will guarantee the film thickness required in the application. Density analyses are commonly run for filling purposes, but they also prove the base fluids used. Elemental and molecular analyses ensure the right base fluids and additives in the lubricant as well as indicating contaminants. Additional analyses can be foaming, demulsibility and air release.

Industrial *gear oils* are quite similar to hydraulic oils, both having only small amounts of additives. However, gear oils contain extreme pressure additives with sulfur (see Chapters 4 and 5). Typically, viscosity and density are measured and elemental and molecular analyses are run. Kinematic viscosity will ensure having the right viscosity and ability to lubricate the gear. Density is run for filling purposes, but will also indicate the proper base fluids in the blend. Elemental and molecular analyses will prove the amount of sulfur in the blend as well as show possible contamination.

Lubricants for *combustion engine oils* contain a higher percentage of additives than hydraulic and gear oils. They contain a lot of dispersants and detergents to keep the combustion engine clean and thereby enhance the life of the lubricant.

Typical analyses are kinematic viscosity at high temperatures (i.e. anticipated operating temperature), density and TBN. In addition, elemental and molecular analyses are performed. The viscosity measurement will ensure the correct quality of the engine oil. TBN will ensure that detergents and dispersants are added. Elemental and molecular analyses will ensure the right composition.

7.4 Used Oil Characterization

Used oils are characterized for different reasons. The analyses will show the history and indicate the present condition of the lubricant. The characterization is primarily performed on lubricants used in real applications [3, 4].

Characterization is done by carrying out analyses on a representative sample from the selected application. The analyses may evaluate either physical or chemical properties (refer to Chapter 2). These analyses will give information about the condition of the lubricant with respect to the amount of additives left, the condition of the base fluid such as oxidation (i.e. base fluid degradation), the presence of wear particles and different types of contaminants.

As mentioned above, the reason that lubricants need to be characterized varies. For instance, a customer may have an identified problem requiring condition analyses to solve the issue. Other reasons may include planned maintenance where knowledge about the lubricant condition is necessary. Regular condition analyses on samples from one application allow trends to be obtained and shown. Planning is equally important to used oil characterization as it is for the characterization of nonused lubricants. It is important to determine which information is needed and to which accuracy and also decide upon the methods to use. This will form the basis for the sample volume needed.

7.4.1 Selection of Analyses

The sample is analysed to determine additive consumption, base fluid degradation, wear and contamination. A selection of analyses is carried out for each type of lubricant based on experience (see Figure 7.7).

Additive consumption during use depends on the required degree of additive action during the specific lubricating conditions. Depending on the lubricating conditions the consumption may be different for different types of additives. The additive content and the level of consumption are evaluated by analysing the amount of typical additive elements in the lubricant.

Typical analyses to prove additive consumption are TBN, elemental analysis and molecular analysis. TBN will indicate if dispersants and detergents are consumed in combustion engine oils, which is likely if the combustion process is incomplete, forming soot. Elemental analysis is useful for investigating the consumption of some additives as well as for observing contamination or wear particles. Typical elements are calcium and magnesium in detergents and phosphorus in antiwear additives. Molecular analysis will show additive consumption such as a reduction in antioxidant concentrations.

Base fluid degradation usually means oxidation, but may also be caused by permanent shear with a reduction in molecular weight of the chemical constituents of the base fluids.

Common analyses are kinematic viscosity and molecular analysis. These analyses are sometimes supplemented with analysing either TAN or TBN. A reduction in molecular weight is shown as a viscosity decrease. However, since a viscosity decrease may also depend on mixing with a lubricant of lower viscosity, further investigations are usually needed to determine permanent shear loss.

Oxidation can be determined by molecular analysis, for example by observing peaks in the FTIR spectrum showing chemical species formed during ageing. Molecular analysis is widely

Information about
- additive consumption
- base fluid degradation
- wear of the contact
- contamination

Figure 7.7 The overall purpose of used oil characterization

used to determine, for example, whether the lubricant is in good condition, contaminated by another lubricant or oxidized. Oxidation may also be identified by a TAN increase or a TBN decrease. Ageing usually results in sludge and thereby a viscosity increase.

Wear occurs on the solid surfaces in the lubricated contact. It can be identified by elemental analysis of wear particles in the lubricant. Detection of metal existing in the contacting bodies will indicate wear. These metals may be, for example, iron, copper, tin or other metals from the component. High concentrations indicate high wear rates where further investigations may be recommended and/or where it may be necessary to find solutions to avoid future wear.

Lubricant contamination may be water, dissolved contaminants and/or particles. Different amounts of *water* may be entrained depending on the base fluid chemistry. Water may occur as free or dissolved water. Mineral-based lubricants may naturally contain 100–200 ppm of water, while ester-based lubricants can contain up to 1000 ppm of water. Esters are more polar (water-like) than mineral-based lubricants and can therefore dissolve more water. Large amounts of water is analysed by distillation and small amounts of water is analysed via titration (e.g. Karl Fisher titration) [4].

Dissolved contaminants may be surface active substances and therefore give rise to foaming, deaeration or demulsifying problems. Analyses of these properties and elemental analyses can be run to demonstrate contamination. For instance, sodium may indicate glycol entrainment in engine oils.

Particles may originate from wear of the contact or may be introduced from the environment or other parts of the system. Particles in the lubricant may give rise to increased wear, catalyse oxidation and hydrolysis, and reduce the cold flow properties by seeding the crystallization of wax molecules. Filtering may reduce particles. The number of particles are counted and reported in three different sizes:[2] >4 mm, <6 mm and >14 mm. Particle counting is difficult with numerous causes of errors. The most basic rule is to have a representative sample collected in a clean container. Particle counting is typically done manually in a light optical microscope or automatically in an optical particle counter.

7.4.2 Analysis Examples of Selected Applications

Different analyses are run for lubricants used in different applications. The selection of analyses is based on experience and knowledge about problems that may arise.

Hydraulic oils are commonly light in colour and, therefore, easy to inspect visually. Viscosity measurements will indicate contamination or ageing. The total acid number (TAN) is usually measured as a supplement to viscosity measurements as an indication of ageing. A TAN of 2 or higher may indicate ageing.

The hydraulic application is very sensitive to contaminants. Even small concentrations of contaminants may affect interfacial properties, such as air and water entrainment. Water entrainment is determined since it may disturb the pressure transfer ability of the hydraulic oil. However, analyses of foaming, demulsibility and air release properties are costly and therefore run only when a clear problem is identified.

Particles can be detrimental to the hydraulic pump. Therefore, hydraulic oils are commonly filtered both during filling and continuously during use. Particle counting is carried out to ensure low levels.

---

[2] According to ISO standard 4406

Lubricant Characterization

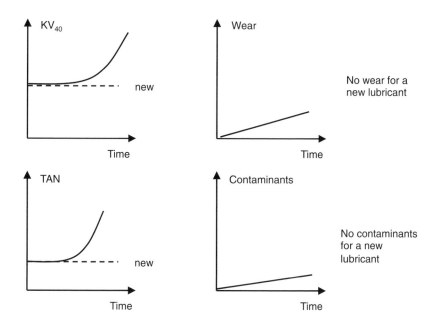

Figure 7.8 Examples of trend curves for a hydraulic oil that has started to age

Trend curves are shown in Figure 7.8. In this case the oil has started to age, which is seen in a TAN increase and a viscosity increase. An increased TAN may cause wear, which is shown as increased metal content. The contaminant level is here shown to slowly increase with usage.

Gear oils are analysed in much the same way as hydraulic oils even though fewer analyses are considered to be required. They are commonly light in colour, allowing visual inspection. Viscosity measurements will indicate good or bad condition, ageing or blending with, for example, another product. Elemental analysis is done to identify additive consumption, for example sulfur content, wear or contamination.

Trend curves for a gear oil are shown in Figure 7.9. In this case the lubricant has been diluted, which is shown as a decrease in viscosity. This viscosity decrease may result in poor lubrication and after some time this leads to catastrophic wear or seizure, which is shown as increased metal content and decreased extreme pressure additive content. Contaminant levels slowly increase, which is expected to be normal during usage.

Combustion engine oils work in a complex system. Therefore, many analyses are needed to pinpoint potential problems. Visual inspection may be difficult due to the usual dark colour. The fact that these oils contain detergents and allow emulsions to form may complicate visual inspection further.

Oil condition characterization usually starts by measuring the viscosity. A decrease in viscosity may mean fuel dilution while an increase in viscosity may indicate oxidation. Fuel dilution may be determined by measuring the fuel content on its own and can be indicated by a decrease in flash point. A clear trend of viscosity, fuel content and flash point will together make it easier to determine fuel dilution. Fuel dilution may cause wear, which can be proven by analysing the additive and the metal content.

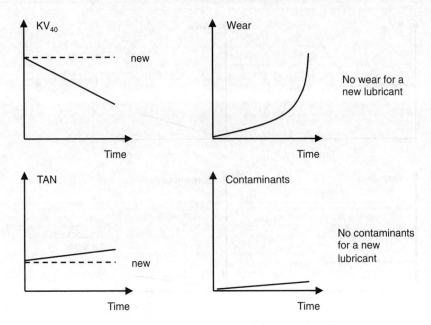

Figure 7.9 Examples of trend curves for a diluted gear oil

Engines have high operating temperatures. This enhances oxidation, which gives rise to the formation of sludge and acidic components. Ageing can be determined by molecular analysis where chemical species formed during ageing can be identified. In combination with molecular analysis, a decrease in TBN may be an indication of oxidation. Cooling of the engine is carried out to prolong the life of both the engine and the lubricant. This is done by glycol, which is completely separated from the lubricant. However, under certain circumstances there may be a leakage of glycol to the engine oil. Even *very small* amounts of glycol may be detrimental and give heavy sludge formation. Therefore, the presence of glycol is important to identify quickly.

Occasionally the combustion is incomplete, giving rise to soot formation leading to soot particles being drawn into the engine oil. The soot is distributed in the lubricant by the dispersants and removed via the oil filter. The soot content will give an indication of the engine oil condition and is therefore measured together with dispersant levels. High soot content is usually found together with a reduced dispersant content.

Trend curves are shown in Figure 7.10. Sample A shows contamination of glycol. Therefore, the viscosity increases rapidly. Sample B shows fuel dilution causing continuous viscosity decrease. At some point the lubricating film thickness will reach a limit where rapid wear occurs. During operation, additives are consumed, which can be demonstrated and shown by a reduction of TBN. Both samples A and B show an expected decrease of TBN. Contamination is expected to increase during usage. The shown contamination curves have no correlation to either glycol contamination or fuel dilution. They are both randomly selected.

7.5 Summary of Used Oil Analyses

A summary for hydraulic oils, gear oils and combustion engine oils is shown in Table 7.1.

Lubricant Characterization

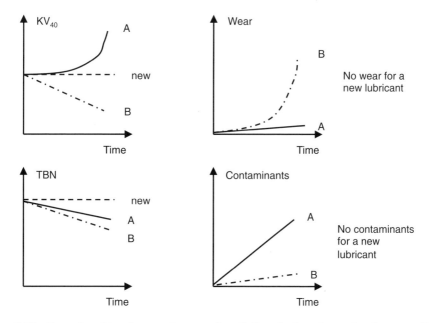

Figure 7.10 Examples of trend curves for an engine oil. Samples have been taken from two different engines. Lubricant A is contaminated by glycol and lubricant B is diluted with fuel

Table 7.1 Summary of used oil analyses performed on hydraulic oils, gear oils and engine oils

Analysis	Hydraulics	Gear	Engine
Appearance	X	X	
KV_{40}	X	X	X
KV_{100}	X		X
TAN	X	X	
TBN			X
Water	X		X
Contamination and metal content	X	X	X
Particle count	X		
Additive elements	X	X	X
Fuel content			X
Flash point			X
Soot content			X
Glycol content			X
Oxidation (ageing)			X

KV_{40}: Kinematic viscosity at 40 °C; KV_{100}: kinematic viscosity at 100 °C; TAN: total acid number; TBN: total base number

References

[1] Fritz, J.S. and Schenk, G.H. (1979) *Quantitative Analytical Chemistry*, 4th edn, Allyn and Bacon, USA.
[2] Torbacke, M. (2001) On the Influence of Mixing and Scaling-Up in Semi-Batch Reaction Crystallization. Doctoral thesis from KTH, Sweden.
[3] Nadkarni, R.A. (1991) *Modern Instrumental Methods of Elemental Analysis of Petroleum Products and Lubricants*, Special Technical Publication, STP 1109, ASTM, USA.
[4] Denis, J., Briant, J. and Hipeaux, J.-C. (2000) *Lubricant Properties: Analysis and Testing*, Editions Technip, Institut Francais du Pétrole, France.

8

Surface Characterization

The tribological contact results in changes of the surfaces, for example in surface composition, microstructure and topography. Good tribological performance of lubricated contacts depends on the formation of beneficial lubricant films or tribofilms (surface films made of new material). Thus, it is of great interest to characterize the surfaces in order to understand the mechanisms of friction, wear and lubrication, both during good function and in the case of failure. In this chapter the methodology for surface characterization is discussed. Commonly used methods are briefly presented, with a focus on what information can be gained.

Four aspects of surface characterization are covered: microscopy of contact surfaces and cross-sections, surface measurement, hardness measurement and chemical analysis of surfaces (see Figures 8.1 and 8.2). In order to characterize friction, wear and lubrication mechanisms, a combination of different methods giving different information is most often required.

Surface microscopy includes high magnification studies of small details of the surface, but also microscopy at lower magnification showing, for example, tribofilm coverage. *Cross-section microscopy* gives information of, for example, thickness and structure of tribofilms, coatings or deformed surface layers.

Methods for *surface measurement* give information of the physical topography of the surface. Surface roughness directly affects the lubricant film parameter and thus also the lubricant film regime. *Hardness measurement*, such as micro hardness and nanoindentation, gives valuable complimentary information about the surfaces. Both surface roughness and surface hardness are important factors influencing friction and wear.

Chemical analysis gives information of the chemical composition of the surface. It can also provide information about chemical structure, adsorbate bonding and molecule bonding. Some techniques also provide the possibility to obtain information from different depths below the surface, that is depth profiles. A large number of analysis techniques exist, all having different benefits and drawbacks. Some of the most useful techniques for characterization of tribological surfaces are presented.

The characterization work can be done at different magnification levels. Methods for different scales of microscopy, surface measurement and hardness measurement are summarized in Table 8.1.

Lubricants: Introduction to Properties and Performance, First Edition.
Marika Torbacke, Åsa Kassman Rudolphi and Elisabet Kassfeldt.
© 2014 John Wiley & Sons, Ltd. Published 2014 by John Wiley & Sons, Ltd.

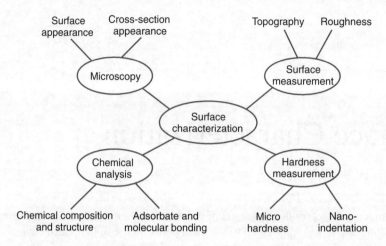

Figure 8.1 Four aspects of surface characterization of tribological surfaces

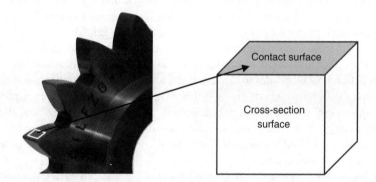

Figure 8.2 Schematic of a piece of material cut out from a component. Both the original contact surface and the cross-section surfaces are of interest to characterize (For a colour version of this figure, see the colour plate section)

Table 8.1 Some methods for microscopy, surface measurement and hardness measurement, ordered by scale of detail in information

Level of detail	Microscopy	Surface measurement	Hardness measurement
1 mm–10 cm	Visual inspection	Stylus profiler	Macro hardness
1 μm–1 mm	Light optical microscopy Optical interference microscopy	Optical interference microscopy	Micro hardness
10 nm–10 μm	Scanning electron microscopy Focused ion beam microscopy Atomic force microscopy	Atomic force microscopy	Nanoindentation
0.1 nm–100 nm	Transmission electron microscopy		

8.1 Surface Characterization of Real Components

Surface characterization of real components can be done as part of, for example, the development, production and failure analysis work. This means that both nonused and used surfaces can be examined.

8.1.1 Examination of Nonused Surfaces

Surfaces of original components are examined before use to ensure functionality during operation. To secure a high quality lubricated contact, it is important that the nonused surfaces meet the requirements. For example, the lubricant film formation and the additive action should be promoted (see also Figure 5.5).

The nonused surfaces are controlled, for example the surfaces are visually inspected and surface roughness and hardness are measured. The manufacturer has defined tolerances for each property and the components used in an application have passed these. If the surface properties are not within the specifications, the surfaces must be further analysed to determine if, for example, the correct materials, manufacturing methods and surface treatments have been used.

8.1.2 Examination of Used Surfaces

Examination of used surfaces is done in order to evaluate the tribological conditions of a functional system, and also to understand and categorize failure. When disassembling a system different parts of the tribological contact are collected, including some of the lubricant (see more about analysis of used oil in Chapter 7). Additional information about the application, for example about running conditions and performance, should also be collected [1].

In order to evaluate the tribological conditions, that is the quality of the lubricated contact, ideally components from different stages of the system life should be collected and characterized, giving a trend analysis of the conditions. If the surfaces are from a system that has reached a state of severe failure, it becomes difficult to retrieve information about normal running conditions. However, since very few disassemble well-functioning systems, this type of information can be difficult to receive from real applications. To some extent relevant information can be collected from field, bench or component testing.

Regardless of component type and the system examined, the following steps may be possible to carry out without destroying the components:

1. Inspect the system visually as it is. Disassemble the system and inspect the components and their surfaces visually. This will give information about, for example, macroscopic damage and deformations, wear tracks, discoloured surfaces, contamination and changes of the lubricant.
2. Use a simple magnifier as a supplement to the visual inspection. For documentation purposes take pictures with a regular camera.
3. Measure the surface topography and roughness, using a portable stylus profiler. This will give information about, for example, the depth of wear tracks and changes in topography.

At this stage, large components have to be cut in order to fit into the different microscopy and analysis instruments. The further examination may include one or several of the following steps and may be done in any order:

4. Measure the micro hardness of the contact surfaces in order to evaluate changes from the original surfaces. Perform nanoindentation testing if very thin films are of interest. Measurements on cross-sections provide information about how the hardness varies with depth due to, for example, deformation and heating.
5. Measure the surface roughness at higher resolution, in order to evaluate the wear and lubrication conditions. When the optical interferometer (e.g. vertical scanning interferometry (VSI)) is used, lubricant that can disturb the reflection of light must be removed. Atomic force microscopy (AFM) provides very detailed information from small surface areas.
6. Use scanning electron microscopy (SEM) to study, for example, wear, wear mechanisms, surface damage, contaminations and tribofilm formation. SEM studies are preferably combined with X-ray spectroscopy (e.g. energy dispersive X-ray spectroscopy (EDS)), giving information of elemental composition. The vacuum environment of the SEM requires dry surfaces; thus the surfaces must be clean from, for example, lubricants.
7. Study cross-sections of the surfaces using, for example, light optical microscopy (LOM) or SEM. Properly prepared and etched cross-sections give information about, for example, how the microstructure changes with depth due to heating, deformation or cracking.
8. Perform chemical analysis of the surfaces. Such analyses give information about composition and chemistry of lubricant films and tribofilms. Depending on what information is required an appropriate method is selected. X-ray photoelectron spectroscopy (XPS), Auger electron spectroscopy (AES), secondary ion mass spectroscopy (SIMS) and infrared spectroscopy (IRS) are methods giving information from the outermost atomic layers of the surface.
9. Perform transmission electron microscopy (TEM). A sample preparation can be made using the focused ion beam (FIB) technique. TEM analysis may give very detailed information about material structure and composition.

8.1.3 Characteristics of Application Examples

Oils are formulated with the aim to prevent wear and reduce friction, typically by forming different types of surface films. How well this aim is fulfilled is shown by the performance of real components in service. It can also be evaluated by used oil characterization and by surface characterization. For all components, it is of interest to identify the active wear mechanisms and to measure the surface roughness and hardness, in order to evaluate possible changes from the original surfaces. In addition, surface characterization can indicate the presence of, for example, corrosion, contamination, misalignments, construction faults and material faults. The characteristics of the selected application examples, hydraulics, gears and combustion engines, are presented below.

Hydraulic oils contain antiwear (AW) additives based primarily on phosphorus chemistry. It is expected that the AW additives form a chemisorbed surface film. It is interesting to study chemical composition, surface coverage and thickness of the film. The presence of phosphorus is used to identify the film. Although the AW tribofilms are expected to protect the surfaces, wear may occur. Adhesive wear is the most typical surface damage mechanism.

Gear oils often contain a lot of sulfur and a reaction between sulfur and iron at the contact surface is expected. It is interesting to study the extent of this reaction, as well as the chemical composition, surface coverage and thickness of the surface film. Here sulfur is used to identify the film. Indications of surface degradation can be permanent deformation of the gear teeth, different mechanisms of sliding wear and corrosion. In the worst cases, the gear may have been subjected to excessively high loads and friction, causing extensive adhesive wear, that is scuffing. This occurs when the lubricant film breaks down due to too high loads and temperature. Surface fatigue due to shear stress concentrations may also cause surface damage, such as micro pitting and even loss of larger pieces of materials.

In *combustion engines* protecting films are also expected to form on the contacting surfaces. There are many different tribological contacts in an engine, and also many different engine oils. In engine oils AW additives are used and surface films including, for example, phosphorus are expected. Since there are many types of tribological contacts and conditions, there may also be different types of wear.

8.2 Microscopy Techniques

In this section, microscopy techniques for revealing surface and cross-section appearances are presented. The *surface appearance* can be examined using various methods, ranging from visual inspection by the naked eye to advanced electron microscopy techniques. In a similar way, *cross-sections* of the surfaces reveal information of, for example, thickness and structure of tribofilms, coatings or deformed surface layers and bulk materials. There are different ways of preparing cross-sections, for example by grinding and polishing samples that are cast in epoxy. To obtain well-defined small-scale cross-sections more site-specific preparation is required. This can be obtained by the focused ion beam (FIB) technique, which is therefore included in this section.

Some terms are often used:

- The *spatial* (or *lateral*) *resolution*; describes how large surface features have to be in order to be visible (i.e. how close two points or lines can be and still be resolved).
- The *highest possible magnification* is limited by the spatial resolution. The better the spatial resolution, the higher is the possible (or meaningful) magnification.
- The *depth of field* describes within what height interval the surface appears to be in focus.

8.2.1 Visual Inspection

The most straightforward way to study a surface is to use the naked eye, a simple magnifying glass or a camera. Visible surface roughness, damage or colour changes may indicate the occurrence of, for example, oxides, tribofilms, deformation or wear. The eye will also identify contamination and extraordinary changes of the lubricant. However, for more detailed tribological studies, microscopy at higher magnification is required.

8.2.2 Light Optical Microscopy (LOM)

The light optical microscope (LOM) is the most common microscopy technique. It is easy to use and can be used in room atmosphere, imaging all surfaces that reflect light. The images also show colours.

The objective lens is the most important part of the microscope. It collects the reflected light from the illuminated surface and forms the image. The best theoretical resolution is about 0.3 μm. It is set by the wavelength of the light and by the performance of the objective lens (i.e. numerical aperture). This indicates that the highest possible magnification is about 1000×. In a conventional LOM the depth of field is very short, and all parts of an uneven or rough surface will not be in focus, but new techniques for optical 3D (three-dimensional) measurement and microscopy are evolving.

In tribology, besides microscopy of the contact surfaces, LOM is often used to study polished cross-sections, revealing microstructural changes by depth. Such changes can be due to, for example, heating, deformation or cracking.

8.2.3 Optical Interference Microscopy

Optical interference microscopes combine an optical microscope and an interferometer into one instrument that can produce 3D images and measurements of the surface [2]. *Vertical scanning interferometry* (VSI) is one method. It is a noncontact measurement method that is performed at room atmosphere. It is easy to use, suitable for studies of many types of tribological surfaces and large surfaces can be measured.

An interferometer is an optical device that splits a beam of light into two separate beams and then combines them. One beam is reflected from the sample surface and the other beam is reflected from a reference mirror. When the beams are recombined they create an interferogram that reflects their difference in travel length, that is the height of the surface.

In VSI, white light is used and the interferogram is used to find the height at each surface position. The best vertical resolution is about 10 Å. The spatial resolution is limited by the optical system (as for LOM) and can theoretically be about 0.3 μm with the best objective lens, but is often limited to about 1 μm.

There are some requirements on the surface. The surface must reflect light, because the beam reflected by the sample surface must be reflected back through the objective lens. Therefore, the surface cannot be completely black. It can also be difficult to measure very rough or porous surfaces, since tilted surfaces will reflect the light away from the lens.

VSI is an effective method for *tribology studies* and is often used as a complement to SEM. VSI is used to image the surfaces and to measure the topography and roughness (see Figure 8.3). Height deviations smaller than 0.1 μm can easily be measured.

8.2.4 Atomic Force Microscopy (AFM)

Atomic force microscopy (AFM) belongs to a family of scanning probe microscopy (SPM) techniques that all give small-scale information [3]. SPM techniques are based on scanning a small tip over the surface, monitoring some kind of interaction between the tip and the surface. With AFM the surface is imaged and the topography is measured with very

Surface Characterization

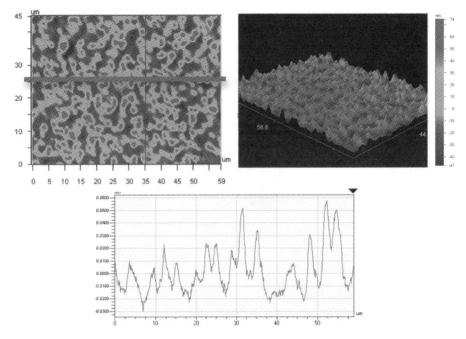

Figure 8.3 Examples of VSI results. The size of the imaged area is 59 μm times 45 μm. Colours (or a grey scale) illustrate the height of the surface in the 2D (upper left) and 3D (upper right) images, the scale ranging from −47 to 74 nm. The position of the line profile (bottom) is indicated in the 2D image (For a colour version of this figure, see the colour plate section)

high resolution (see Figure 8.4). It is performed at room atmosphere and all materials can be analysed.

With AFM the surface is mechanically examined using a very sensitive tip mounted on the end of a flexible cantilever. The tip is scanned across the surface while recording the very small attractive or repulsive force between the outmost atoms of the tip and the surface atoms. The force magnitude depends on the tip–sample distance. A piezoelectric scanner provides the scanning motion. A laser beam is used to monitor the movement of the cantilever caused by the tip–sample interaction.

At its best, the AFM has a vertical resolution of about 1 Å and a spatial resolution of 20 Å. The resolution is mainly set by the tip, which typically has an end radius of 2 to 20 nm. Only small and rather smooth surface areas can be measured, since standard scanners typically have less than 5–10 μm of vertical range and less than 100 μm of horizontal range.

For *tribology studies*, AFM may provide extraordinary information of, for example, mild wear mechanisms or details of thin tribofilms.

8.2.5 *Scanning Electron Microscopy (SEM)*

The scanning electron microscope is the most widely used type of electron microscope. It is the prime technique for imaging and examining wear and damage mechanisms. The most important features of SEM are high spatial resolution and large depth of field, but also the

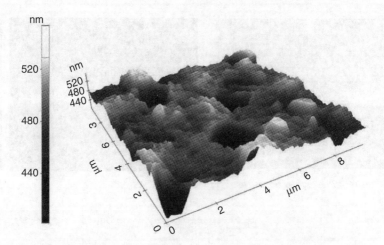

Figure 8.4 AFM image of a part of the surface shown in Figure 8.3. The size of the imaged area is 10 μm times 10 μm. The height varies between 400 and 520 nm

possibility to combine imaging with elemental analysis using EDS (see Section 8.5.4). For a comprehensive description of SEM, see Reference [4].

A finely focused electron beam is scanned over the surface to be imaged. The incident electrons, with energy typically of 5–20 keV, interact with the sample atoms within a few microns of the surface. This interaction generates various signals emitting from the surface, for example electrons and X-ray photons. The SEM image is formed from the intensity of electron emissions from the surface. When the electron beam is scanned over the surface, the intensity of emitted electrons from each position point of the surface sets the intensity of the corresponding point of the image (see Figure 8.5).

There are different modes of SEM imaging. Most SEM pictures found in publications are showing a topographical contrast that reveals surface structure. Due to the large depth of field a three-dimensional appearance of the surface may be obtained (see Figure 8.6). Compositional contrast is an imaging mode where surface areas of different elemental composition give different contrasts. Areas of light elements show a dark contrast, while areas of heavy elements show a bright contrast (see Figure 8.7).

With the best instruments the spatial resolution can be about 10 Å,[1] but is often about 50 Å or more, permitting magnification up to about 100 000×. The resolution is limited by the size of the electron beam, which depends on the type and design of the electron source and the lens system. It is possible to continuously alter between the overview and details, that is to go from low (<100×) to high magnification.

SEM is made in vacuum, which means that the samples must be clean and dry. Electrically insulating surfaces, which get negatively charged by the electron irradiation, can be analysed by SEM after, for example, being coated with a thin layer of a gold alloy. However, the development of advanced SEM systems is fast and today most samples can be studied with SEM techniques.

---

[1] The unit Ångström, 1 Å = 0.1 nm = 10^{-10} m

Surface Characterization

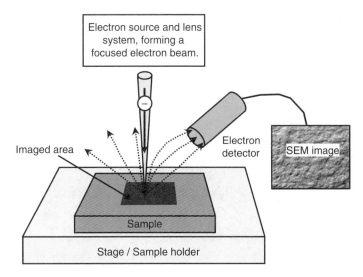

Figure 8.5 Principle of SEM. The focused electron beam is scanned over the surface (scanned area = imaged area). The electron detector counts the emitted electrons. The intensity of emitted electrons sets the contrast of the SEM image

SEM for *tribology studies* is diverse and very effective for examining wear and damage mechanisms. By combining the SEM with EDS elemental information can also be obtained. Many SEM instruments allow for relatively large and heavy samples (approximately up to 1 kg in weight and 10 cm in size). Thus, it is not always necessary to cut the component into pieces. SEM with EDS is also used for particle analysis during used oil characterization, providing information about size, shape and composition of the particles.

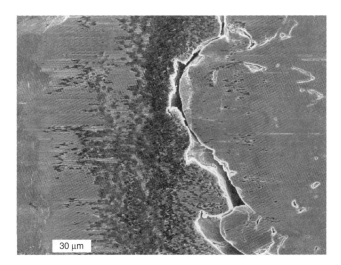

Figure 8.6 SEM image of a surface from a tribological contact, showing cracking and surface contamination that gives the dark contrast

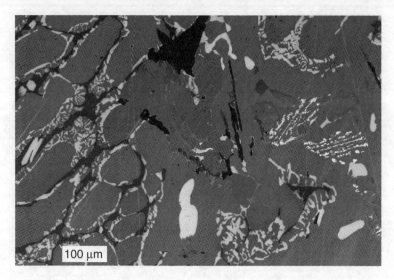

Figure 8.7 SEM image illustrating compositional contrast. The microstructure of a polished surface of a metallic alloy is shown. The dark areas are rich in carbon and the light areas are rich in molybdenum

8.2.6 Focused Ion Beam (FIB)

Focused ion beam (FIB) is a technique used to analyse and manipulate a small surface area of particular interest using a well-controlled beam of gallium (Ga) ions [5]. It is usually integrated with SEM, making it possible to select a surface area of interest and to position the ion beam more accurately.

FIB has proven very useful for tribology studies. It is, for example, used for preparing cross-section surfaces for further SEM studies and for preparing cross-section samples for TEM (see Section 8.2.7).

The procedure to prepare a sample starts by using SEM to select a surface area of interest (see Figure 8.8). A thin coating is deposited on the surface in order to protect it from the ion

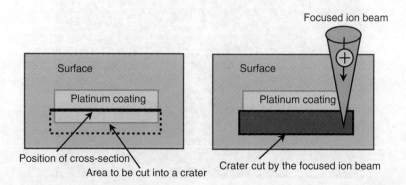

Figure 8.8 Schematic of the FIB method for preparation of cross-sections and TEM samples. The position of the cross-section is selected from the SEM image. A protective coating is deposited on the surface. The focused ion beam is used to cut a crater into the surface

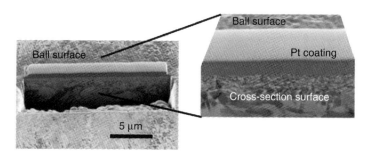

Figure 8.9 FIB prepared cross-section of the surface of a steel ball from a ball-on-disc test. The length of the crater is about 15 μm and the thickness of the protective coating (Pt) is about 1 μm. The close-up shows a fine-grained microstructure close to the surface

bombardment during etching. The focused ion beam is then used for etching a crater into the surface. For TEM analysis a sample is cut as a thin slice of the surface. An FIB cross-section for SEM studies of a tribology surface is shown in Figure 8.9.

8.2.7 Transmission Electron Microscopy (TEM)

Transmission electron microscopy (TEM) provides high-resolution imaging and analysis [6]. Both imaging at atomic resolution and chemical analysis of a very small sample volume are possible. The capability of imaging and analysis of fine details makes it the major analysis method in many scientific fields, from biology to physics and engineering. Depending on the configuration of the instrument, different kinds of imaging and analysis can be performed, providing a large variety of information about the sample.

In a conventional TEM a thin specimen (<100 nm in thickness) is irradiated with an electron beam (see Figure 8.10). The electrons interact with the specimen when passing through the sample. When the electrons exit the specimen, they may be unaffected, they may have lost some energy or they may have altered direction. The TEM image contrast is caused by variations in absorption and diffraction of electrons, for example due to differences in thickness, composition and crystal orientation. A lens system provides the formation of an image of the specimen captured on a screen. The TEM instrument is also often equipped with an EDS detector for elemental analysis.

It can be difficult to prepare samples for TEM. It is particularly difficult to prepare cross-section samples of surfaces. TEM is a relatively complicated and expensive technique to use and should primarily be selected when detailed information is needed.

For *tribology studies* TEM provides detailed information about, for example, crystal structure, grain size and elemental composition, which may be crucial for the fundamental understanding of, for example, lubricant film formation and friction mechanisms. An example of a TEM cross-section of a thin tribofilm formed on a coating is shown in Figure 8.11 [7]. After sample preparation using the FIB, the sample was moved to the TEM instrument.

8.3 Surface Measurement

Surface roughness is an important parameter affecting the tribological contact. It directly affects the lubricant film parameter (see Equation (1.5)), but also the friction and wear of the

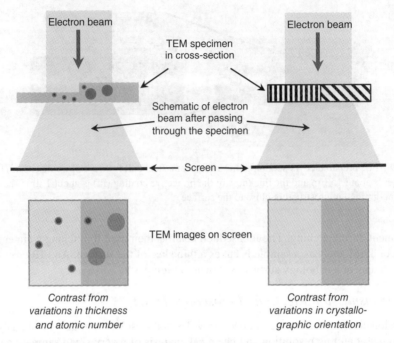

Figure 8.10 Schematic of the principle of TEM imaging. An electron beam irradiates a thin specimen. Interactions between the electrons and the specimen give rise to different contrast mechanisms, due to variations in, for example, absorption and scattering

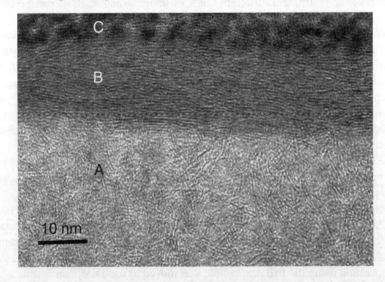

Figure 8.11 High resolution TEM image of a cross-section of a coating (MoSeC) after ball-on-disc testing. Area A shows the original amorphous atomic structure of the coating. Area B shows a deformed top layer of the coating, where the atomic structure has been lined up forming a low friction material. The upper part of the figure, area C, is influenced by the protective coating from the sample preparation by FIB

Surface Characterization

lubricated contact. The *surface measurement* data that quantitatively describe a surface can be expressed by different statistical parameters. Such parameters can be utilized, for example, for describing and comparing nonused and used surfaces.

8.3.1 Statistical Surface Parameters

No surface is perfectly flat; for example all manufacturing processes leave machining marks at some scale. The *surface topography* can be divided into form, waviness and roughness. Form is the large-scale shape, for example a cylinder. Waviness is a superimposed structure, for example from the machine tool during manufacturing of the cylinder. Roughness is the superimposed topography at the micro or submicron scale.

Statistical parameters are defined using surface measurement data [2]. These parameters include the commonly used R_a and R_q parameters. R_a is the average roughness, that is the arithmetic mean of the magnitude of the deviation from the mean line, and R_q is the root mean square roughness. The R_a and R_q parameters are defined as

$$R_a = \frac{1}{n} \sum_{i=1}^{n} \left(z_i - \bar{z}\right) \tag{8.1}$$

$$R_q = \sqrt{\frac{1}{n} \sum_{i=1}^{n} \left(z_i - \bar{z}\right)^2} \tag{8.2}$$

where $\bar{z}$ is the mean height of the surface, z_i is the surface height at position i and n is the number of positions (see Figure 8.12). For all positions from $i = 1$ to $i = n$, the difference between the height and the mean height is measured. To get the R_a value the summary of these differences is divided by the number of positions.

The R_a and R_q parameters are general guidelines of surface roughness. However, they typically prove too general to describe a roughness profile. For example, surfaces with sharp spikes or deep pits may have the same R_a value (see Figure 8.13). Other parameters exist that can be more appropriate.

It is important to note that different measurement techniques and different operating conditions using the same technique will give different results. Thus, one should be careful when using these parameters and when comparing surfaces on the basis of parameter values.

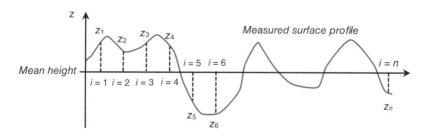

Figure 8.12 Schematic description of z_i values ($i = 0, \ldots, n$) along a measured surface profile

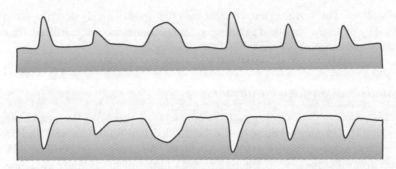

Figure 8.13 Example of two different surface profiles having the same R_a values

8.3.2 Contacting Stylus Profiler

Most surfaces, including contaminated surfaces, can be measured with a contacting stylus profiler. The surface profile is measured by moving a diamond stylus or a glass sphere along the surface [2]. Since the stylus is touching the surface there may be a risk of deforming the surface.

There are many types of stylus profilers covering a range of testing conditions. It is possible to measure height differences between 10 nm and 10 mm and lengths from 1 to 100 mm. Diamond styluses have radii of 50 nm to 0.5 μm and glass spheres have radii of 1 to 500 μm. The load can be between 0.5 and 100 mg. The spatial and vertical resolutions depend on the size and radius of the stylus. The smaller the stylus the better is the resolution, since a smaller stylus follows the surface more closely (see Figure 8.14).

Stylus profilers are easy to use and offer good price performance. Several parallel scans can be combined to obtain a 3D measurement.

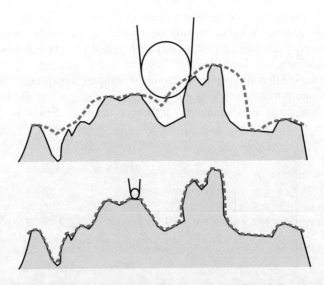

Figure 8.14 Schematic of how two different stylus profilers follow the surface

8.3.3 Microscopy Techniques

The VSI and AFM microscopy techniques presented in Section 8.2 also provide surface measurement data. VSI is very useful for typical tribological surfaces. It produces 3D measurements of the surfaces in a very rapid and easy way. AFM is an outstanding technique for high resolution imaging. It produces very precise 3D measurements of surface areas of limited size and height.

8.4 Hardness Measurement

The surface hardness of a component affects the lubrication conditions in a tribological contact. Hardness describes the material's resistance to plastic deformation. The hardness of the surfaces and possible tribofilms are often different from those of the bulk material and are therefore of great importance for understanding the tribological contact. It is a mechanical parameter that is relatively easy to measure. A sharp indenter is forced into the surface under a controlled load. The size of the resulting or residual indentation, after removing the indenter, is related to a hardness number.

8.4.1 Macro and Micro Hardness

In order to get the hardness value of a bulk material a *macro hardness* test is used, for example the Rockwell or Brinell tests. Here a hard spherical indenter (several mm in diameter) and a high load (tens of kg) are used.

As a rule of thumb, to prevent influence from surrounding materials the indentation depth should be 1/10 of the depth of interest for the measurement. In order to get information from the surface region rather than from the bulk material, *micro hardness* techniques must be applied. The most commonly used micro hardness tests are those of Vickers and Knoop. Both use small diamond indenters of pyramidal geometry where the load is kept below 1000 g. The Knoop and the Vickers hardness numbers are designated by HK and HV respectively (e.g. HV_{20} means Vickers hardness measured at 20 g load).

8.4.2 Nanoindentation

In *nanoindentation* very small loads and indentation tips are used. Thus, the information depth is also very small, often less than 100 nm. The indentation load is recorded versus the indentation depth (see Figure 8.15). The load is higher during loading than during unloading. The test gives information about hardness, but also about the elastic modulus of the surface [8]. The elastic modulus is a measure of how much a material deforms elastically, or strains, under stress.

Modern instruments may be equipped with AFM for positioning and analysis and with small tips for scratching at low loads. Such instruments can be used for mechanical characterization of very precisely selected areas of, for example, thin tribofilms.

8.5 Surface Analysis Techniques

Surface analysis gives information about the composition and chemistry of the surface material. In this section some methods having the widest application in tribology will be briefly

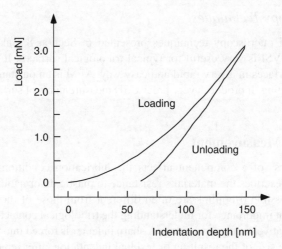

Figure 8.15 Applied load versus indentation depth (displacement) during loading and unloading in a nanoindentation test

presented. The presentation focus on the information retrieved, the characteristic performance parameters and the sample requirements [9–12].

All surface analysis methods are typically based on the following steps:

- The surface is stimulated, for example, by some irradiation or other kind of stress.
- Interaction processes with the surface material result in a response from the surface.
- A spectrum is created from the detected signal.

For the selected methods, the surface is stimulated with a beam of particles or photons, that is electromagnetic radiation. The particles can be electrons or ions. The electromagnetic radiation can be infrared (IR) light or X-ray radiation (see Figure 8.16). The surface responds by sending out electrons, ions or photons. These signals are detected and give information about the surface.

8.5.1 Selected Methods

There are many methods available for surface analysis. Some methods relevant for tribological surface analysis have been chosen for a closer description. These are X-ray photoelectron spectroscopy (XPS), Auger electron spectroscopy (AES), time-of-flight secondary ion mass spectroscopy (ToF-SIMS), Fourier transform infrared (FTIR) spectroscopy and X-ray spectroscopy (EDS or EDX). They are presented in Table 8.2 with the surface stimuli used and the type of detected signal.

XPS, AES and ToF-SIMS are methods that give information from the outermost part of the surface ($\approx$1–20 atom layers). FTIR spectroscopy is useful for probing the bonding of adsorbed atoms and molecules to a surface. It is a versatile method and can be used for analysis of liquids

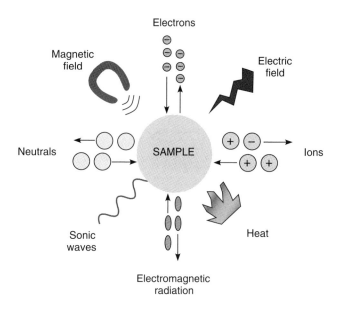

Figure 8.16 Different stimuli of the surface are shown together with possible responses that may result. Interesting information, a signal, is possible to filter out and detect, giving an analysis method

(refer to Chapter 7) and gases. EDS is included because of its usefulness in combination with SEM and TEM.

8.5.2 Analysis Performance Parameters and Terminology

Every surface analysis method has its benefits and limitations. The method to use for a specific analysis problem depends on one or more performance parameters, that is on what kind of information that can be provided.

Table 8.2 Selected surface analysis methods, with type of beam used for irradiation and type of signal detected

	Irradiation by		
Detection of	Electrons	Ions	Photons
Electrons	AES (Auger electrons detected)		XPS (X-ray beam, photo electrons detected)
Ions		ToF-SIMS (secondary ions detected)	
Photons	EDS (in SEM and TEM, X-ray photons detected)		FTIR (beam of IR, reflected IR photons detected)

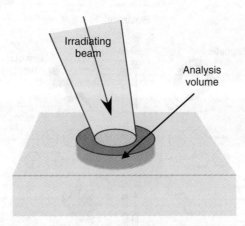

Figure 8.17 Schematic of the analysis volume. The surface is irradiated by a particle or photon beam. The detected response originates from the visualized analysis volume. This volume is often wider than the irradiated area and also has a certain depth

The analysis volume, from which the information is obtained is basically set by the size of the analysis area and the information depth. The analysis area can be larger than the spot size of the irradiating beam, due to spreading of the beam below the surface. Different beams have different penetration depths, depending on the type of beam and material. Interaction between the beam and the material occurs in the whole penetration volume. However, the information depth can be smaller than the penetration depth of the beam. This depends on the possibility for the detected signal to emit from the surface.

Different surface analysis methods provide different levels of information. Some methods give only *elemental information*, that is information of what kind of atoms or elements are present. Other methods also give *chemical information*, that is information about how the atoms are bonded to each other. Some methods give *molecular information*, for example on how specific molecules are adsorbed to a surface.

The *information depth* describes from which maximum depth the detected signal originates, while the *depth resolution* gives information on how well the depth from where it originates can be decided. The information depth required depends on the surface characteristics of interest to study (see Figure 8.18). For example, it may be interesting to study:

- Physisorbed or chemisorbed molecules, which require information from the outmost atom layer or some atom layers (<10).
- Thin lubricant films or tribofilms, for example a chemically reacted layer or a layer of deposited material, which requires information from a thickness of about 100 nm.
- Surface deformations and transformations, or coatings, requiring information from depths of some microns or even more.

The *lateral resolution* describes how well positions of the surface can be determined. The requirement of the lateral resolution depends on how small surface features should be analysed, for example wear particles or other particles embedded in the surface, or small variations (colour, structure, etc.) in surface films.

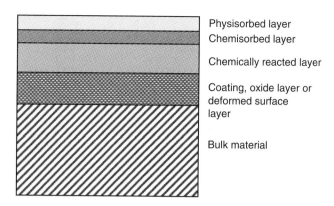

Figure 8.18 Schematic of a cross-section surface revealing possible types of surface layers. Analysis methods are selected depending on the depth of interest

The *detection limit* is the smallest amount of an element or species that must be available in the material to be possible to measure. In addition, only some methods are able to detect hydrogen or helium. This is important to consider since lubricants contain mainly hydrocarbons.

If a *composition quantification* is requested, some methods can provide quantified results directly, while others need a set of reference surfaces to get fairly good results.

The *atmosphere requirements* differ a lot. Some analyses can be made at room atmosphere, while the most surface-sensitive methods require *ultra-high vacuum* (UHV) to avoid the influence of adhered contaminants. Sample requirements also differ. Some methods can be used to analyse any material while other methods may require, for example, electrically conductive materials.

8.5.3 Depth Profiling and Chemical Mapping

Knowing the concentration profile versus depth is valuable both for nonused and used surfaces. AES, XPS and ToF-SIMS are suitable methods for depth profiling using ion sputtering. An ion beam, typically argon (Ar) ions, is scanned over the surface. The energy from the ions is transferred to the surface atoms. These will be sputtered atom layer by atom layer from the surface (see Figure 8.19). By switching between analysing and sputtering, a depth profile is obtained. The intensity of the signal from different elements or species is measured as a function of sputtering time. These data are then transferred into concentration as a function of depth.

Depth profiles can reveal the composition of, for example, reacted layers and deposited films, as a function of depth. Maximum profiling depth by ion sputtering is about 1 µm.

With some methods it is be possible to obtain a chemistry map of the surface. The analysis beam is scanned over the surface and the intensities from different elements or chemistries are measured at different positions of the surface. The collected information is then used to create elemental or chemistry maps (see Figure 8.20).

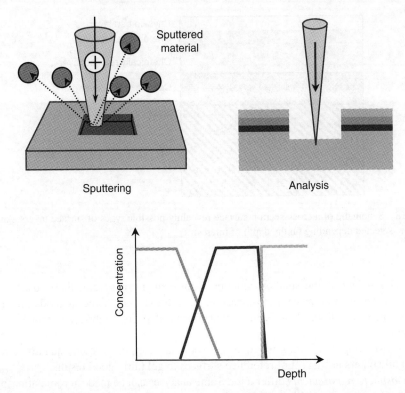

Figure 8.19 Schematic of depth profiling using ion sputtering. An ion beam is used to gradually remove surface atoms by sputtering. The surface is analysed at different depths. A depth profile showing the concentration versus depth is obtained (bottom)

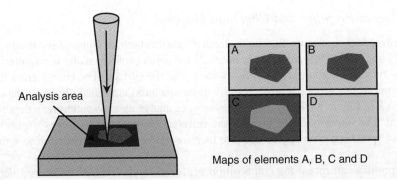

Figure 8.20 Schematic of chemical mapping. The surface is analysed at different positions. The maps show the presence of selected elements or chemistries (light contrast). Elements A and B are found in the same area, element C in the other area and element D in the whole area

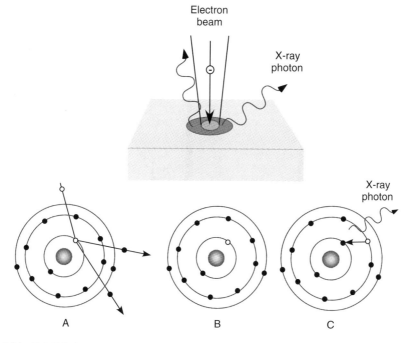

Figure 8.21 In EDS the surface is irradiated with an electron beam. The interaction process includes: an electron from the beam knocks out an inner shell electron (A), an inner shell vacancy is created (B), an electron from a higher shell is filling the vacancy and an X-ray photon is emitted (C)

8.5.4 Energy Dispersive X-Ray Spectroscopy (EDS)

In X-ray spectroscopy characteristic X-rays are used to identify the elements of a sample [4]. One benefit of energy dispersive X-ray spectroscopy (EDS) analysis, which is a technique to detect X-ray photons, is the ease of use in combination with SEM and TEM. The characteristic X-rays are generated from the interaction process between the electron beam and the sample atoms. EDS provides elemental information, but no information about bonding.

The sample is irradiated with electrons with energy high enough to ionize the atom (see Figure 8.21). The ionized atom has a vacancy in an inner shell. An electron from another shell fills this inner shell vacancy. Then the energy difference between these two electron positions will be released in the form of an X-ray photon (or as an Auger electron, see Section 8.5.5).

SEM and TEM instruments are often equipped with an EDS detector that collects and measures the energy of individual X-ray photons emitted from the sample. The energies of the emitted X-ray photons are characteristic since the electron binding energies of different elements are characteristic[2] (see Figure 8.22). Thus, we can identify the chemical elements of the sample from the energy of the emitted characteristic X-rays.

---

[2] X-ray nomenclature: K indicates an initial K shell vacancy and α and β that this vacancy has been filled with an electron from the L shell and the M shell respectively. L indicates an initial L shell vacancy.

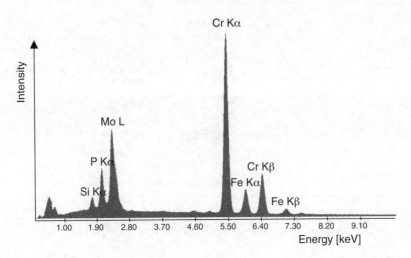

Figure 8.22 EDS spectrum of the grey phase shown in Figure 8.7. The spectrum shows intensity versus energy of the detected X-ray. The peaks clearly identify the presence of Cr, Fe, Mo, P and Si, but also of B, C, O and N at low energy

EDS analysis in SEM can be made from large areas (using a scanning electron beam) or from small areas (using a fixed electron beam). Elemental mapping is also possible. The incoming electrons will spread into a volume below the surface due to scattering. The X-ray photons are generated and emitted from this volume, which is typically about a micron in both depth and width depending on the electron energy and material. Thus, EDS gives information from a larger depth than, for example, XPS and AES.

For *EDS analysis in TEM*, the analysis volume can be kept very small since the TEM sample is very thin. Therefore, the lateral resolution is better for EDS in TEM than in SEM. In TEM the analysis depth equals the sample thickness.

For *tribology studies* EDS in SEM is a very commonly used analysis method. This is mainly due to the combination with SEM, giving the possibility to combine good imaging with elemental analysis. EDS in SEM may be more suitable for surface deformation and wear mechanism studies than for studies of very thin lubricant films. This is due to the information depth that is typically about 0.3–1 μm. The information depth becomes smaller by using lower energy of the electron beam, which may allow for analysis of thin lubricant films or tribofilms. Another benefit of using lower electron energy is the sensitivity increase to light elements (e.g. carbon, nitrogen and oxygen).

8.5.5 *Auger Electron Spectroscopy (AES)*

Auger electron spectroscopy (AES) is a useful method for analysing tribological surfaces [12, 13]. The Auger electrons are generated from the interaction process between the electron beam and the sample atoms. AES is used for analysis of very small surface features, for high resolution elemental mapping and for depth profiling using ion sputtering. Most AES instruments are equipped with an electron detector for SEM imaging. Thus, the surfaces can be imaged using the SEM function and the electron beam can be positioned very precisely for the AES analysis.

Samples must be electrically conductive to be analysed since the electron beam will cause an electric charge-up of electrically insulating surfaces. The samples must also be clean and dry since the analysis is performed in ultra-high vacuum (UHV). All elements, except for hydrogen (H) and helium (He) can be detected.

The surface is irradiated with a focused electron beam (as in EDS in SEM). Auger electrons are emitted from the surface when the incoming electrons interact with the sample atoms. An incoming electron knocks out an inner shell electron, thus creating an inner shell vacancy (see Figure 8.23). The following relaxation of the atom includes the process where an electron from an outer shell moves to the inner shell vacancy. The excess energy is released as an Auger electron (or as an X-ray photon, as detected in EDS). The emitted Auger electrons are detected and their energies are measured. The energy of the Auger electron depends on the binding energies of the electrons involved in the Auger process (denoted 1, 2

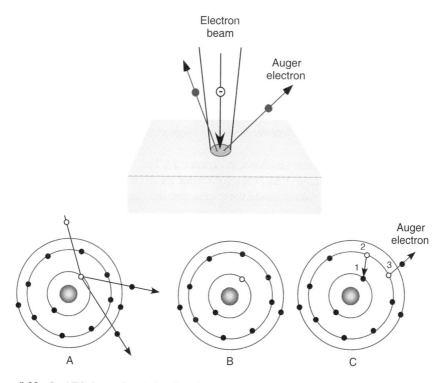

Figure 8.23 In AES the surface is irradiated with an electron beam. The Auger process includes: an electron from the beam knocks out an inner shell electron (A), an inner shell vacancy is created (B), an electron from a higher shell is filling the vacancy and an Auger electron is emitted (C)

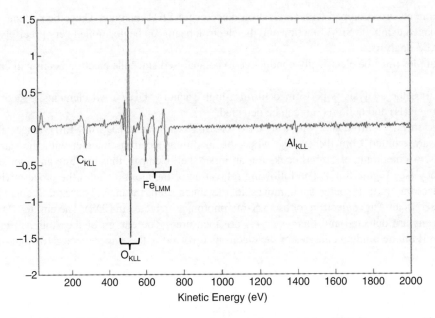

Figure 8.24 AES differentiated spectrum from a surface containing C, O, Fe and some Al

and 3 in the figure).[3] These are characteristic for different elements. Thus, the energies of the emitted Auger electrons are characteristic for the element. In addition, information about chemical bonding can be obtained. This is revealed as changes of the peak energy and shape, reflecting the slight changes in electron binding energies due to the chemical bonding of the atom.

An example of an AES spectrum is shown in Figure 8.24. The Auger electron signal is small and superimposed on a continuous background of secondary electrons. Therefore, Auger peaks are more easily detected by differentiating the energy distribution. Thus, a differentiated AES spectrum is often presented.[4]

An important benefit of AES is the very small analysis volume. The information depth is very narrow, as only about 20 atom layers are analysed. This is because only the Auger electrons generated from near the surface can travel to the surface and escape from it, without losing any energy. The very good lateral resolution, less than 20 nm, is due to the electron beam that can be focused into a very small spot.

For *tribology studies*, AES is most useful for analysis of very small features in the surface and for thin surface film. In addition, it is also useful for high resolution chemical mapping of the surfaces and for depth profiling by sputtering (see Figure 8.25).

---

[3] The Auger nomenclature: KLL means the process with an initial vacancy in the K shell, filled by an electron from an L shell and the emission of an electron from an L shell (as illustrated in Figure 8.23); LMM means the process with an initial vacancy in the L shell, filled by an electron from an M shell and the emission of an electron from an M shell.
[4] The differentiated spectrum shows $dN(E)/dE$ versus E, where $N(E)$ is the number of electrons of energy E.

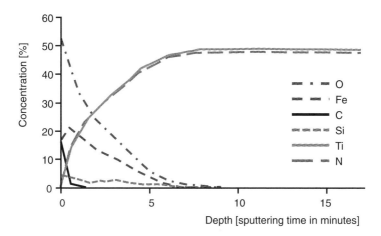

Figure 8.25 AES depth profile of a titanium nitride (TiN) coated surface that has been sliding against a steel material, thus showing mainly Fe and O on the surface. The high C content on the surface is contamination

8.5.6 *X-Ray Photoelectron Spectroscopy (XPS)*

X-ray photoelectron spectroscopy (XPS) and *electron spectroscopy for chemical analysis* (ESCA) are two names for the same technique. Both names occur in the literature and they can be used interchangeably [12, 14]. XPS is one of the most widely used surface analysis methods. It is used for analysis of thin surface layers, giving information about both elements and chemical bonding. The information depth is about 20 atom layers. XPS is also used for depth profiling using ion sputtering.

The samples must be clean and dry since the analysis is performed in ultra-high vacuum (UHV). All types of materials that can withstand the UHV can be analysed. All elements except for hydrogen (H) and helium (He) can be detected.

In XPS the surface is irradiated with X-ray photons (see Figure 8.26). An X-ray photon interacts with an atom and an inner shell electron (by the photoelectric effect). This means that an electron, referred to as the photoelectron, is emitted from the atom.[5] The X-ray photon energy is well defined. Thus, the kinetic energy of the photoelectron gives a direct measurement of its original binding energy to the atom. Measurement of the kinetic energy of the photoelectrons permits the elemental identification by the relation

$$E_{photoelectron} = E_{X\text{-}ray} - E_{binding\ energy}$$

where $E_{photoelectron}$, $E_{X\text{-}ray}$ and $E_{binding\ energy}$ are the energies of the detected photoelectron, the stimuli X-ray photon and the original electron bonding respectively. Typically, the X-ray beam has $E_{X\text{-}ray} = 1486.6$ eV.[6]

---

[5] The XPS nomenclature: a given electron energy state is characterized by four quantum numbers (n, s, l, j). The XPS notations 1s and 2s mean that a photoelectron from the l = 0 (i.e. j = $1/2$) subshell from the K (n = 1) and L (n = 2) shells respectively.

[6] When energetic electrons irradiate Al metal, characteristic X-ray radiation from Al is generated. The X-ray of a specified energy, here AlKα, may be filtered using a monochromator. The energy of AlKα = 1486.6 eV.

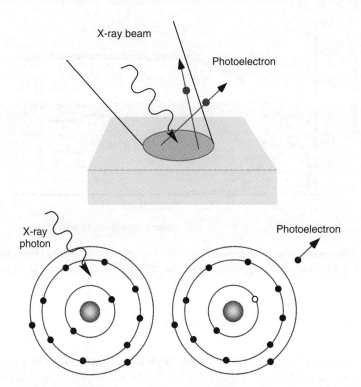

Figure 8.26 In XPS the surface is irradiated with X-ray photons. The interaction process includes absorption of the X-ray photon and emission of a photoelectron

An important benefit of XPS is the small information depth. The X-ray photons penetrate deep into the sample surface. However, only photoelectrons generated from near the surface can travel to the surface and escape from it without losing any energy (as for the Auger electrons in AES). Therefore, only photoelectrons generated within about the top 20 atom layers are bringing valuable information. The lateral resolution can be as good as 10 μm, but is often much larger.

An XPS spectrum of an aluminium oxide surface that has been sliding against PTFE (Teflon) is shown in Figure 8.27. The spectrum shows signals from the transferred PTFE layer and from surface contamination. An XPS spectrum also shows Auger electron peaks, here exemplified by the KLL peak from fluorine, denoted F_{KLL}.

There is a shift in the electron binding energy when an atom is bonded to atoms of other elements in a compound. In XPS the photoelectron energy is detected with a precision good enough to separate different compounds. In Figure 8.27 the C1s peak is clearly separated into two peaks, one corresponding to carbon bonded to carbon and the other to carbon bonded to fluorine, as in PTFE. Such energy shifts, called chemical shifts, form the basis for the chemical information. It is, for example, possible to differentiate between iron in a metallic state or bound as, for example, an oxide, sulfide or chloride.

For *tribology studies* XPS is very useful for identifying the chemical composition and structure of thin surface films, for example thin tribofilms, lubricant films or other reaction

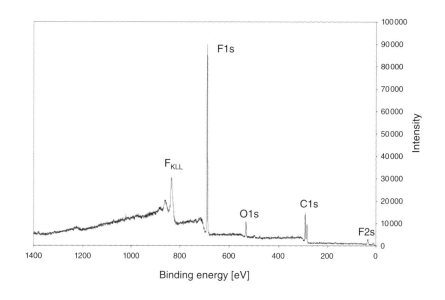

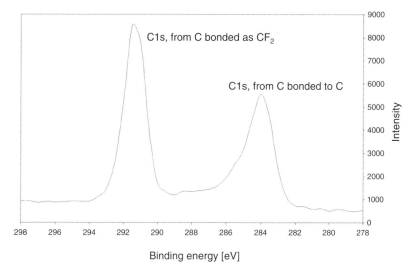

Figure 8.27 XPS spectrum of thin PTFE transfer film on an aluminium oxide surface, survey spectrum (upper) and detail spectrum of the two C1s peaks (lower). PTFE gives rise to peaks from C and F, that is C1s, F1s, F2s and F_{KLL} peaks. Surface contamination gives rise to peaks from O and C, that is O1s and C1s. Due to a chemical shift the C1s peak is separated into two peaks

layers [15]. XPS was used to verify the reaction between tungsten (W) from a coating and sulfur (S) from a lubricant additive (see Figure 8.28). The XPS spectrum shows that W appears both as WS_2 in the tribofilm and as WC in the coating. This is revealed as a small difference in binding energy of the photoelectrons originating from W in the two different compounds. The result indicates that a low friction WS_2 tribofilm has been formed on the steel surface that was sliding against the coated surface [16].

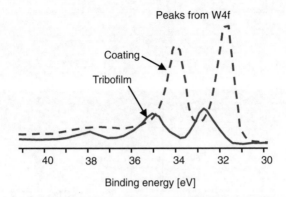

Figure 8.28 XPS spectra from two surfaces, showing the two W4f peaks (5/2 and 7/2) in detail. The peaks from the tribofilm are shifted towards a higher binding energy in relation to the peaks from the coating

8.5.7 Secondary Ion Mass Spectroscopy (SIMS)

Secondary ion mass spectroscopy (SIMS) provides a very detailed chemical analysis of the surface [17]. It is possible to detect low concentrations of all elements including hydrogen. It is very surface-sensitive, allowing for detection of the top one or two atomic layers. A relatively narrow spot can be used, allowing for analysis and chemical mapping of small surface features.

The analysis is performed in UHV. Therefore, it is important to handle the surfaces carefully prior to analysis since the top layers are investigated. All types of materials that can withstand the UHV can be analysed.

The surface is bombarded with a beam of ions (i.e. primary ions). The energy is transferred to the atoms of the sample surface. This results in sputtering of atoms, atom clusters, molecules or molecular fragments (see Figure 8.29). Some of the sputtered particles are ionized, that is negatively or positively charged. These are the secondary ions, which are separated and analysed by a mass spectrometer. Thus, in SIMS the masses of the secondary ions are examined.

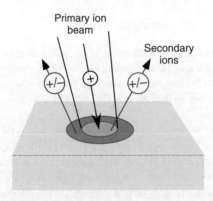

Figure 8.29 In SIMS the surface is irradiated with an ion beam (primary ions), for example gallium ions (Ga). As a result secondary ions (ionized atoms or fragments) are emitted from the surface

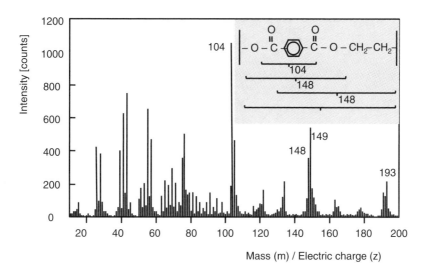

Figure 8.30 Positive ToF-SIMS spectrum (positively charged secondary ions) of the polymer material polyethylene terephthalate (PET). The PET molecule and the masses of some different fragments of it are also shown. All peaks correspond to different fragments of the molecule

There are different modes and types of SIMS operation and mass detection systems. Here *time-of-flight secondary ion mass spectroscopy* (ToF-SIMS[7]) will be covered since it is the most widely used mass detection technique for static SIMS. This is a technique particularly useful for analysis of tribological surfaces.

A ToF-SIMS spectrum shows the detected intensities of secondary ions versus their mass-to-charge ratio (m/z), where mass is in atomic mass units. The mass resolution is very good, $m/\Delta m$ can be as high as 10 000, where m is the measured mass and Δm is the resolution of the measurement of the mass.

The spectrum can be either that of positive ions or negative ions; often both types are presented. Individual ions, ion clusters and molecular fragments can be identified (see Figure 8.30). This spectrum illustrates the large number of possible ionized fragments that are sputtered from the surface. They range from small fragments consisting of only one or two atoms to large fragments consisting of large parts of the polymer structure. This fact often makes interpretation of the spectra complicated due to too much and complex chemical information.

Depth profiles may be obtained by alternating sputtering and analysis. The prime ion gun is used for the analysis. The same ion gun can also be used for sputtering, although some ToF-SIMS instruments are equipped with an additional ion gun for sputtering.

For *tribology studies* ToF-SIMS can be used to obtain detailed chemical information of the very outmost atom layers of the surfaces. In addition, relatively small surface features can be analysed. Another benefit of ToF-SIMS is that hydrogen can be detected, allowing for investigations of hydrocarbons in surface films formed in lubricated tribological contacts.

---

[7] The ToF analyser is briefly described: Short pulses of primary ions are irradiating the surface, resulting in secondary ions emitting the surface. The secondary ions are all given a specific kinetic energy and their flight time for a specific flight distance is measured. Ions with different masses will have different flight times, while heavier masses will travel more slowly. Thus the mass/charge ratio of the ions can be derived.

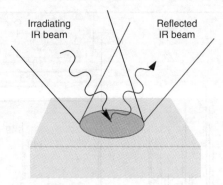

Figure 8.31 In FTIR spectroscopy the sample is irradiated with IR light of different amounts of energy. The intensity of the reflected or transmitted IR light is measured

8.5.8 Fourier Transform Infrared Spectroscopy

Fourier transform infrared (FTIR) spectroscopy is the most widely used method within a family of vibration spectroscopy methods. It is a method to study molecules by examining the interaction between infrared (IR) light and molecular vibrations [18]. It can provide chemical information from adsorbed molecules and additive films on surfaces. FTIR spectroscopy can be used to examine liquids and gases.

Depending on the structure of a molecule, it will vibrate at different frequencies. IR light is electromagnetic radiation (photons) in a certain energy range (or frequency range). When a molecule is irradiated by IR light, a particular photon frequency may match the vibrational frequency of one or some of the atom bonds of the molecule. For FTIR analysis of solid surfaces, a reflectance technique must be applied. Very surface-sensitive techniques are available. Using the reflectance technique the surface is irradiated with IR light of different energy and the intensity of the reflected light is measured (see Figure 8.31). The spectrum shows low intensity at energies that have been absorbed, that is peaks corresponding to bonds present in the surface (see Figure 8.32). The Fourier transform method is used to obtain an IR spectrum in a whole range of energies simultaneously.

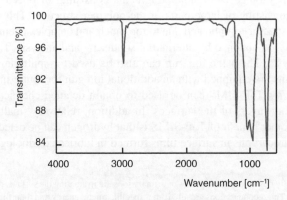

Figure 8.32 Example of an FTIR spectrum. Wavenumber corresponds to frequency or energy of the IR light

For tribology studies, FTIR spectroscopy can be used for analysis of, for example, organic compounds of lubricant films and adsorbates bonded to the surfaces. FTIR spectroscopy is also used for characterization of lubricants (see Chapters 2 and 7).

8.6 Summary of Surface Characterization Methods

8.6.1 Microscopy and Surface Measurement

A comparison between the microscopy and surface measurement techniques reveals some similarities and differences, as shown in Table 8.3. AFM is the technique providing surface measurements with the best resolution. The drawback is that the surfaces must be relatively smooth and that only relatively small samples are possible to measure. VSI is a fast and easy-to-use technique for most surfaces. The drawback is the relatively poor lateral resolution. SEM is the best technique for surface imaging. The drawback is that no quantified height information is provided.

8.6.2 Surface Analysis

Performance parameters for the presented analysis techniques are summarized in Table 8.4, some of which are further discussed in this section (see also Section 8.5.2).

The *lateral resolution* is influenced by the width of the used irradiating beam and the interaction between the beam and the sample. It can also be set by resolution of a position-sensitive detector. A comparison between the presented methods is shown in Figure 8.33. Techniques using an electron beam for irradiation show the best lateral resolution.

Small features require methods probing small analysis areas and volumes. AES combines a small analysis area with good surface sensitivity, that is very small surface features can be analysed. SIMS may also be useful. Particles or surface films of about a micron or larger may also be analysed with EDS in the SEM.

Using analysis techniques in TEM, for example EDS, may give excellent spatial resolution. This requires preparation of TEM samples, which may be difficult and time consuming. Also AES in combination with the FIB technique for well-defined etching of, for example, cross-section of surface features provides an excellent set-up for analysis at high spatial resolution. However, such instrument set-ups are still quite rare.

The *information depth* depends either on how deep below the surface the irradiating beam reaches, that is at which depth the signal can be generated or from which maximum depth the generated signal can travel to get out from the surface. The information depth and lateral resolution of the presented methods are compared in a map shown in Figure 8.34.

FTIR techniques are useful for probing *bonding* of chemisorbed atoms and molecules to a surface and can thus be used for analysis of adsorbed molecules and reacted surfaces layers. SIMS can also provide information on how the atoms of the outmost atom layers are bonded. For chemical analysis of reacted layers SIMS and XPS can be useful.

To reveal the *chemical composition* of tribofilms, surface-sensitive methods that can be combined with depth profiling are useful. Again SIMS and XPS are good choices. If the films have varying thickness and/or are unevenly distributed over the surface, for example due to rough surfaces, AES may be a good choice. With AES a depth profile over a small surface area can be obtained, but the chemical information will not be as detailed as with XPS or SIMS.

Table 8.3 Summary of performance parameters of the presented microscopy and surface measurement methods. All values are approximate and may depend on operating conditions and the analysed material

Performance parameter	LOM	VSI	AFM	SEM	Stylus profiler
Information	2D (3D) imaging (no height measurement)	3D topography and imaging	3D topography and imaging	3D imaging (no height measurement)	2D profile (may be combined to 3D)
Best lateral resolution	0.3 μm	0.3 μm (often 1 μm)	20 Å (often 100 Å)	10 Å (often 50 Å)	<nm – μm
Best vertical resolution	Small depth of field at high magnification	10 Å	1 Å	Large depth of field	<nm – μm
Magnification	10–2000×	10–2000×	500–10^6×	20–10^6×	(Line profile: <100 mm)
Contact force	No contact	No contact	≪0.1 mg	No contact	0.5–80 mg
Comments	• All materials • Several contrast mechanisms in advanced instruments	• All materials • The surface must reflect light • Not too rough or porous surfaces	• All materials • Small and smooth areas • Possible to measure wet surfaces and in liquid	• Difficult with insulating or magnetic samples • Vacuum compatible samples • Several contrast mechanisms • Advanced instruments for difficult samples	• All materials. • Risk for damage of soft surfaces. • Different types, using different stylus sizes and loads, giving different performance.

Table 8.4 Summary of performance parameters of the presented surface analysis methods. All values are approximate and depend on instrument, operating conditions and the analysed material

Performance parameter	EDS in SEM	EDS in TEM	AES	XPS	ToF-SIMS	FTIR
Chemical information	Elemental	Elemental	Elemental and some chemical bonding	Elemental and chemical bonding	Elemental and chemical bonding	Molecular bonding
Lateral resolution	≈ 0.5–1 μm	< 0.01 μm	0.02 μm	10 μm	0.2 μm	10 μm
Information depth	≈ 0.5–1 μm	Sample thickness (<100 nm)	< 5 nm	< 5 nm	1–3 monolayers	<nm – μm (reflective)
Quantification	Good	Good, but small volume	Good, using standard	Good, using standard	Poor	Poor
Depth profile method	No	No	Ion sputtering	Ion sputtering	Ion sputtering	No
Sample requirements	• Conductive materials (or use thin coating) • Not H or He	• Special TEM sample preparation • Not H or He	• Conductive materials • Not H or He	• All materials • Not H or He	• All materials • All elements	• Organic compounds, molecular groups.
Vacuum requirements	Medium	Medium	UHV	UHV	UHV	UHV – Normal atmosphere
Main benefits for tribological evaluation	• Easy-to-use with SEM • Good quantification	• Very small analysis volume	• Very good lateral resolution • Surface sensitive	• Surface sensitive • Chemical structure • All materials	• Very surface sensitive • Chemical structure • Very low detection limit • Detects H	• Adsorbates on surfaces • Hydrocarbon molecular groups • Some techniques are very surface sensitive

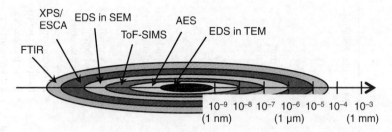

Figure 8.33 Illustration of lateral resolution of the presented surface analysis methods

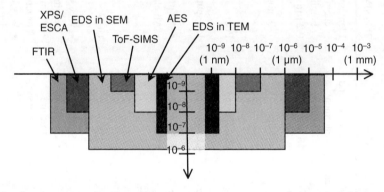

Figure 8.34 Illustration of information depth and lateral resolution of the presented surface analysis methods

The *detection limit* depends on the relation between the intensity of the detected signal and the background or noise. It also depends on the size of the analysis volume, or the total number of atoms in the analysis volume. ToF-SIMS has a very low detection limit. The ability to detect H or He also differs, where EDS, AES and XPS cannot be used. FTIR spectrometry or SIMS are useful for studying adhered or reacted layers originating from the lubricant since lubricants contain mainly hydrocarbons.

References

[1] van Beek, A. (2006) *Advanced Engineering Design – Lifetime Performance and Reliability*, TU Delft.
[2] Thomas, T.R. (1998) *Rough Surfaces*, 2nd edn, Imperial College Press.
[3] Leggett, G.J. (2009) Scanning probe microscopy, in *Surface Analysis: The Principal Techniques*, 2nd edn (eds J.C. Vickerman and I.S. Gilmore), John Wiley & Sons.
[4] Goldstein, J. (ed.) (2003) *Scanning Electron Microscopy and X-Ray Microanalysis*, 3rd edn, Springer.
[5] Giannuzzi, L.A. and Stevie, F.A. (eds) (2005) *Introduction to Focused Ion Beams*, Springer.
[6] Willims, D.B. and Carter, C.B. (2009) *Transmission Electron Microscopy – A Textbook for Materials Science*, 2nd edn, Springer.
[7] Gustavsson, F., Jacobson, S., Cavaleiro, A. and Polar, T. (2013) Frictional behaviour of self-adaptive nanostructural Mo–Se–C coatings in different sliding conditions. *Wear*, **303**, 286–296.
[8] Oliver, W.C. and Pharr, G.M. (1992) An improved technique for determining hardness and elastic modulus using load and displacement sensing indentation experiments. *Journal of Materials Research*, **7**, 1564–1583.

[9] Vickerman, J.C. and Gilmore, I.S. (eds) (2009) *Surface Analysis: The Principal Techniques*, 2nd edn, John Wiley & Sons.
[10] Buber, H., and Jenett, H. (eds) (2002) *Surface and Thin Film Analysis*, John Wiley & Sons.
[11] Leng, W. (2008) *Materials Characterization, Introduction to Microscopic and Spectroscopic Methods*, John Wiley & Sons.
[12] Briggs, D. and Seah, M.P. (1990) *Practical Surface Analysis, Auger and X-Ray Photoelectron Spectroscopy*, 2nd edn, vol. 1, John Wiley & Sons.
[13] Mathieu, H.J. (2009) Auger electron spectroscopy, in *Surface Analysis: The Principal Techniques*, 2nd edn (eds J.C. Vickerman and I.S. Gilmore), John Wiley & Sons.
[14] Ratner, B.D. and Castner, D.G. (2009) Electron spectroscopy for chemical analysis, in *Surface Analysis: The Principal Techniques*, 2nd edn (eds J.C. Vickerman and I.S. Gilmore), John Wiley & Sons.
[15] Naveira Suárez, A. (2011) The Behaviour of Antiwear Additives in Lubricated Rolling–Sliding Contacts. Doctoral thesis from Luleå Universtiy of Technology, Sweden, ISBN 978-91-7439-212-8.
[16] Stavlid, N. (2006) On the Formation of Low-Friction Tribofilms in Me-DLC – Steel Sliding Contact. Doctoral thesis from Uppsala University, Sweden, ISBN 91-554-6743-1.
[17] Vickerman, J.C. (2009) Molecular surface mass spectrometry by SIMS, in *Surface Analysis: The Principal Techniques*, 2nd edn (eds J.C. Vickerman and I.S. Gilmore), John Wiley & Sons.
[18] Pemble, M.E. and Gardner, P. (2009) Vibrational spectroscopy from surfaces, in *Surface Analysis: The Principal Techniques*, 2nd edn (eds J.C. Vickerman and I.S. Gilmore), John Wiley & Sons.

Index

α-value, 25
Abrasive wear
    2-body abrasive wear, 12
    3-body abrasive wear, 13
Absolute value, 137
Accuracy, 137
Acidic component, 34, 86
Activation energy, 35
Additive activation, 77
Additive competition, 78
Additive consumption, 143
Additive film, 15
Additive solubility, 70, 88
Additive types
    antioxidants (AO), 36, 87
    antiwear additives (AW), 75
    corrosion inhibitors, 73
    defoamers, 79
    demulsifiers, 80
    detergents, 85
    dispersants, 84
    emulsifiers, 80
    extreme pressure additives (EP), 76
    friction modifiers (FM), 75
    pour point depressants (PPD), 82
    viscosity modifiers (VM), 81
Additives
    action mechanism, 71
    additive activation, 77
    additive competition, 78
    additive consumption, 143, 145
    additive element, 139, 143, 147
    additive exploration, 71

additive film, 15
additive package, 88, 95
additive solubility, 53, 70, 88
bulk active additives, 71, 81, 85
surface active additives, 69, 73, 78
Adhesive wear, 12, 15
Adsorption, 68, 73, 78
Adsorption layer, 71
Aeration, 30
Air entrainment, 67, 79
Air release, 29, 139
Antioxidant (AO)
    hydroperoxide decomposer, 87
    metal deactivator, 87
    radical scavenger, 87
Antioxidant remains, 139
Antiwear additive (AW), 75
Apparent contact area, 6
Application requirement, 38, 88, 93
Applications
    cam follower contact, 116, 131
    combustion engines, 107, 131, 147
    gears, 104, 129, 147
    hydraulics, 101, 129, 147
    piston ring–cylinder liner contact, 107, 129
    piston–ball joint contact, 129
Applied load, 9, 99, 116
Archard's wear model, 14
Aromatic structure, 47, 55
Arrhenius equation, 35
Asperities, 8, 11
Atomic force microscopy (AFM), 154, 180
Auger electron spectroscopy (AES), 170, 181

Lubricants: Introduction to Properties and Performance, First Edition.
Marika Torbacke, Åsa Kassman Rudolphi and Elisabet Kassfeldt.
© 2014 John Wiley & Sons, Ltd. Published 2014 by John Wiley & Sons, Ltd.

β-value, 23
Base fluid
    base fluid degradation, 142
    base fluid price, 50
    base fluid properties, 45
    diester, 58
    gas-to-liquids (GTL), 55
    monoester, 58
    naphthenic base oil, 53
    natural ester, 57
    nonpolar base fluid, 64, 98
    paraffinic base oil, 53
    phosphate ester, 59
    polar base fluid, 98
    polyalkylene glycol, 59
    polyalphaolefin (PAO), 54
    polyisobutene, 59
    polyolester, 58
    rapeseed oil, *see* Natural ester
    re-refined base oils, 56
    silicone oils, 59
    sunflower oil, *see* Natural ester
    synthesized base fluid, 54
    synthetic esters, 57
    vegetable oils, *see* Natural ester
    very high viscosity index base oil (VHVI), 54
    white oils, 54
Base fluid degradation, 142
Base fluid group
    group I, 48, 60
    group II, 48, 60
    group III, 48, 60
    group IV, 48, 60
    group V, 48, 60
Base fluid properties, 45
Base oil, *see* Base fluid
Bench test, 106, 114, 128
Bioaccumulative, 42, 95
Biodegradability, 42, 95
Boundary lubrication, 11, 76, 121
Bulk active additive, 71, 81, 85

Cam follower contact, 116, 131
Change interval, 103, 110
Characteristic X-rays, 169
Chemical adsorption, 69
Chemical analysis
    chemical composition, 152, 163, 179
    cross-section analysis, 152, 160, 167
    depth analysis, 152, 167
    lubricant analysis, 139
    surface analysis, 163, 165, 179
Chemical corrosion, 74
Chemical mapping, 167
Chemical potential, 66
Chemical reaction, 63
Chemical reaction rate, 35
Chemisorption, 70
Cleveland open cup (COC) method, 28
Coalescence, 29, 32, 80
Coefficient of friction, 8, 79, 121
Cold cranking simulator (CCS), 43, 109
Cold flow properties, 47, 82, 144
Combustion engines
    combustion engine oil, 142, 146
    combustion engine oil formulation, 108, 110
Component test, 113, 128
Conformal contact, 7, 115, 129
Contact area
    apparent contact area, 6
    distributed contact area, 6, 116, 121
    line contact area, 6
    near-contact area, 77
    off-contact area, 77
    point contact area, 6, 115, 123
    real contact area, 6
Contact geometry
    conformal contact, 7, 115
    nonconformal contact, 7, 115
Contamination
    dissolved contaminants, 56, 144
    dissolved water, 32, 144
    fuel dilution, 145
    glycol contamination, 146
    particle counting, 135, 139, 144
    water entrainment, 29, 138, 144
Conventional base fluids, *see* Base fluids
Corrosion inhibition, 37, 73
Corrosion inhibitor, 73
Covalent bonding, 64
Cross-section microscopy, 149, 152, 158
Crude oil, 50

Defoamer, 79
Defoaming, 29
Demulsibility, 32, 139
Demulsifier, 80
Density, 23
Depth of field, 153
Depth profile, 167

Desorption, 69, 75
Detection limit, 167, 182
Detergent
  neutral detergent, 86
  overbased detergent, 86
  phenate, 86
  salicylate, 86
  sulfonate, 86
Dewaxing, 51
Diffusion, 67
Diffusivity, 67
Dipole–dipole interactions, 66, 98
Dispersant, 84
Distillation at atmospheric pressure, 51
Distributed contact area, 6, 116, 121
Dry lubricant, 16
Dynamic viscosity, 21, 43, 109

Ecolabel, 42
Elastic deformation, 12
Elastohydrodynamic lubrication, 12, 117
Elastomer compatibility, 47
Elastomer material
  elastomer compatibility, 47
  fluorocarbon rubber (FKM), 97
  hydrogenated nitrile butadiene rubber (HNBR), 97
  nitrile butadiene rubber (NBR), 97
Electrochemical corrosion, 74
Electron binding energy, 172, 174
Electron microscopy, see SEM and TEM
Electron spectroscopy for chemical analysis (ESCA), see XPS
Elemental analysis, 143, 156, 170
Elemental mapping, 170
Emission legislation, 3, 108
Emulsifier, 80
Emulsion, 32, 80, 145
Energy dispersive X-ray spectroscopy (EDS), 157, 169, 181
Environmental conditions, 5, 116, 121
Environmental impact, 40, 95
Environmental properties, 19, 40, 96
Environmentally adapted lubricant, 40, 48, 56
Erosive wear, 4, 12
Ester, see base fluid
Extreme pressure additive (EP), 76

Fick's equation, 67
Field test, 113, 128

Film forming additive, 73
Film parameter, 10, 79, 99
Film thickness measurement
  electrical method, 117
  optical interferometry method, 118
Filterability, 84, 97, 139
Flash point, 27, 139
Fluorocarbon rubber (FKM), 97
Foam build-up, 31, 79, 103
Foaming, 30, 139
Focused ion beam microscopy (FIB), 158, 160
Fourier transform infrared spectroscopy (FTIR)
  FTIR for lubricant analysis, 43, 139
  FTIR for surface analysis, 178, 181
Fretting wear, 13
Friction
  coefficient of friction, 8, 79, 121
  dry friction, 8
  dynamic friction, 124
  friction force, 8, 121
  friction loss, 5
  static friction, 8
  viscous friction, 8
Friction coefficient, see coefficient of friction
Friction modifier (FM), 75
Fuel, 107
Full film lubrication, 4, 11, 99
FZG test, 129

Gas–liquid surface, 67
Gas-to-liquid (GTL) base fluid, 48, 55
Gears
  FZG test, 129
  gear oil formulation, 106
  gear test, 106
  gear tooth contact, 105
Grease, 16

Hardness
  Brinell hardness, 163
  hardness measurement, 163
  Knoop hardness, 163
  nanoindentation, 150, 163
  Rockwell hardness, 163
  Vickers hardness, 163
Heat capacity (specific), 15, 33, 43
Heavy duty engine, 108, 110
High quality contact, 99
High temperature high shear (HTHS), 43, 109

Hydraulics, 101
  hydraulic oil formulation, 102
  hydraulic system, 101
Hydrocarbons, 36, 45
Hydrofinishing, 52
Hydrogenated nitrile butadiene rubber (HNBR), 97
Hydrolytic stability, 37
Hydroperoxide decomposer, 87
Hydrorefining, 51

Inductively coupled plasma (ICP), 43, 139
Infrared (IR) microscopy, *see* FTIR
Inherent viscosity, 24, 93
Interacting surfaces, 3
Interaction forces
  covalent bonding, 64, 70, 101
  dipole–dipole interactions, 65, 98
  ionic bonding, 64, 98
  van der Waal forces, 64, 98
Ionic bonding, 64, 98
ISO VG, 103, 106

Kinematic viscosity, 22, 43, 139

Lambda value (Λ-value), 10
Langmuir adsorption isotherm, 69
Lateral resolution
  microscopy, 153
  surface analysis, 166
Light optical microscopy (LOM), 154, 180
Line contact area, 7
Liquid–liquid surface, 66, 79
Long life properties, 19, 33, 95
Longevity, 19, 87
Lubricant additives, *see* Additives
Lubricant analysis, *see* Lubricant characterization
Lubricant characterization
  absolute value, 137
  accuracy, 137
  air entrainment, 67, 79
  air release, 29, 139
  corrosion inhibition properties, 73
  demulsibility, 32, 139
  environmental properties, 19, 40, 96
  flash point, 27, 139
  foaming, 30, 139
  hydrolytic stability, 19, 37
  oxidation stability, 19, 35
  planning, 133

pour point, 82, 139
reference value, 136
sampling, 135
thermal properties, 32
total acid number (TAN), 34, 139
total base number (TBN), 35, 139
trend analysis, 137, 151
viscosity, 20, 138
water entrainment, 29
Lubricant condition, 139
Lubricant consumption, 56
Lubricant contamination, 144
Lubricant film parameter, *see* Film parameter
Lubricant film thickness, 11, 99, 117
Lubricant formulation, 93
Lubricant properties, 19
Lubricant purposes, 14
Lubricating regime
  boundary lubrication, 11, 76
  elastohydrodynamic lubrication, 12, 117
  film parameter, Λ-value, 10, 79, 99
  full film lubrication, 11, 79, 99
  mixed lubrication, 11, 75, 99
  Stribeck curve, 11
Lubrication regime, *see* Lubricating regime

Magnification, 149, 153, 180
Mainline product, 93
Mass transfer, 67
Material compatibility, 96
Microscopy
  comparison, 179
  cross-section appearance, 149, 152
  depth of field, 153
  lateral resolution, 153
  magnification, 149, 153, 180
  spatial resolution, 153
  surface appearance, 149, 152
Microscopy techniques
  atomic force microscopy (AFM), 154, 180
  focused ion beam microscopy (FIB), 158
  light optical microscopy (LOM), 154, 180
  optical interference microscopy, 154
  scanning electron microscopy (SEM), 155, 180
  transmission electron microscopy (TEM), 159
  vertical scanning interferometry (VSI), 154, 180
Mini-rotary viscometer (MRV), 43, 109
Miscibility, 97

Mixed lubrication, 11, 79, 98
Mixing
   blending, 134
   mixing intensity, 134
   mixing theory, 134
   semi-circular flow, 134
Model test, 115
Molecular analysis, 139
Monograde, 109
Motion type
   combined sliding and rolling, 7
   rolling, 7
   sliding, 7
Multigrade, 109

Nanoindentation, 163
Naphthenic base oil, 53
Natural ester, 57
Near-contact area, 77
Newton's law, 21
Newtonian fluid, 22
Nitrile butadiene rubber (NBR), 97
Noack volatility, 27
Nonconformal contact, 6
Nonconventional base fluids, see Base fluids
Non-Newtonian fluid
   pseudoplastic, 25
   thixotropic, 26
Nonpolar, 64, 98
Nonpolar base fluid, see Base fluid
Nonrepresentative sample, 135
Nonused lubricant characterization, 140

Off-contact area, 77
Oil additives, see Additives
Oil analyses, see Lubricant characterization
Operating conditions, 5, 121, 180
Optical interference microscopy (e.g. VSI), 154, 180
Oxidation
   carboxylic acid, 36, 87
   initiation step, 87
   oxidation process, 35
   oxidation stability, 35
   oxygen consumption, 35
   peroxide radical, 87
   propagation step, 87
   radical, 35, 87
   radical–radical reaction, 87
   termination step, 87

Paraffinic base oil, 53
Particle count, 139, 144
Permanent dipole, 98
Permanent shear loss, 58, 143
Physical adsorption, 69
Physical properties, 20, 139
Physisorption, 69
Pin-on-disc tribotest, 121
Piston ring–cylinder liner contact, 129
Piston–ball joint contact, 129
Plastic deformation, 8, 163
Plastic material, 96
Point contact area, 6
Polar base fluid, see Base fluid
Polar moiety, 69
Polarity, 71
Polyalkylene glycol, 59
Polyalphaolefin (PAO), 54
Polyisobutene, 59
Polymeric materials, 96
Pour point, 27, 139
Pour point depressant (PPD), 82
Premium product, 95
Pump test, 103
Pumpability, 21, 109
Purpose of lubricants, see Lubricant purposes

Qualitative analysis, 133
Quantitative analysis, 133

R_a-value, 161
Re-refined, 49, 56, 60
Real contact area, 6
Reciprocating motion, 116, 123, 131
Reciprocating tribotest, 123, 129, 131
Reference value, 136
Refining process, 50
Renewability, 20, 40, 95
Repeatability, 136
Representative sample, 135, 142, 144
Reproducibility, 136
Requirement specification, 96
Rolling contact, 10, 13
Rolling motion, 7, 104
Rotary tribotest, 128
R_q-value, 161
Running-in, 8, 13, 121

Sample dilution, 35, 133, 136
Sample preparation TEM, 159, 181

Sampling, 133, 136, 141
Scanning electron microscopy (SEM), 155, 180
Scanning probe microscopy (SPM), 154, 180
Sealing material, 41, 96 103
Secondary ion mass spectroscopy (SIMS), 177, 181
Semi-solid lubricant, 16
Shear rate, 21, 25, 76
Shear stability, 26, 43, 81
Shear stress, 10, 21, 153
Sliding contact, 10, 15, 121
Sliding motion, 7, 77, 121
Sliding velocity, 117, 121
Sludge, 35, 84, 138
Sludge formation, 35, 110, 146
Soft metal, 39, 75, 87
Soft metal corrosion, 39, 75, 87
Solid lubricant, 16
Solid–liquid surface, 67, 73
Solid–gas–liquid surface, 66
Solubility, 53, 70, 88
Solvency, 46, 54
Solvent extraction, 51, 56
Spatial resolution, *see* Lateral resolution
Specification, 96, 102
Statistical surface parameters, 161
Steel corrosion, 37, 74
Stokes equation, 30
Stribeck curve, 11
Stylus profiler, 162, 180
Substances of very high concern, 42
Sulfated ash phosphorus sulphur (SAPS), 109
Sulfur, 48, 65, 71
Surface active, 31, 70, 72
Surface active additive, 70, 78, 88
Surface active molecule, 31, 63, 69
Surface activity, 46
Surface analysis
    chemical mapping, 167
    comparison, 179, 181
    depth profiling, 167
    detection limit, 167
    information depth, 166
    parameters, 165, 179
Surface analysis techniques
    Auger electron spectroscopy (AES), 170, 181
    electron spectroscopy for chemical analysis (ESCA), 173, 181

    energy dispersive X-ray spectroscopy (EDS), 169, 181
    Fourier transform infrared spectroscopy (FTIR), 178, 181
    secondary ion mass spectroscopy (SIMS), 176, 181
    X-ray photoelectron spectroscopy (XPS), 173, 181
Surface characterization
    characteristics of selected applications, 152
    examination of non-used surfaces, 151
    examination of used surfaces, 151
    general aspects, 149
Surface composition, 70, 149
Surface fatigue, 13, 101, 153
Surface measurement
    arithmetic mean roughness, R_a, 161
    resolution, 153, 179
    root mean square roughness, R_q, 161
Surface microscopy, *see* Microscopy
Surface roughness, 5, 159
Surface smoothening, *see* Running-in
Synthetic ester, *see* Base fluid

Thermal conductivity, 33
Time-of-flight secondary ion mass spectroscopy (ToF-SIMS), *see* Secondary ion mass spectroscopy (SIMS)
Topography, *see* Surface roughness
Total acid number (TAN), 34, 139
Total base number (TBN), 35, 139
Toxicity, 40, 95
Transmission electron microscopy (TEM), 159
Trend analysis, 137, 151
Trend curve, 145
Tribochemical wear, 13
Tribofilm, 117, 176
Tribological contact, 5, 98, 114
Tribological contact conditions
    local conditions, 8, 115
    local pressure, 8, 12
    local temperature, 8, 15, 106
Tribological test methods
    film thickness measurement, 117
    pin-on-disc tribotest, 121
    reciprocating tribotest, 123
    rotary tribotest, 128
    twin disc tribotest, 124

Tribological testing
    bench test, 113
    component test, 113
    field test, 113
    model test, 115
    planning, 114
    reciprocating motion, 116, 123
    strategy, 115
    unidirectional motion, 116
Twin disc tribotest, 124

Used lubricant characterization, 142
Used oil, 40, 142

Vacuum distillation, 51
van der Waal forces, 64, 98
Vapour pressure, 27, 59
Vegetable oil, see Base fluid
Vertical resolution, 180
Very high viscosity index (VHVI) base oil, 54
Vickers hardness, 163
Viscometer, capillary, 23
Viscometer, rotational, 22
Viscosity
    capillary viscometer, 23
    dynamic viscosity, 22
    kinematic viscosity, 22
    rotational viscometer, 22
    viscosity grade, 103, 106
    viscosity index, 24, 81
    viscosity–pressure dependence, 24
    viscosity–shear rate dependence, 25
    viscosity–temperature dependence, 23
Viscosity grade, 103, 106
Viscosity index, 24, 81
Viscosity index improver, see Viscosity modifier (VM)

Viscosity modifier (VM)
    olefin copolymer (OCP), 82
    polyisobutene (PIB), 82
    polymethacrylate (PMA), 82
Viscosity–pressure dependence, 24
Viscosity–shear rate dependence, 25
Viscosity–temperature dependence, 23
Visual inspection, 138, 153
Volatility, 27

Water content, 139
Water entrainment, 29
Wear
    Archard's wear model, 14
    failure, 13
    mild wear, 13
    running-in, 13
    severe wear, 13
    wear life, 5
    wear rate, 13
Wear mechanisms
    abrasive wear, 12
    adhesive wear, 12
    erosive wear, 13
    fretting wear, 13
    surface fatigue, 13
    tribochemical wear, 13
White oil, 49, 54, 93

X-ray fluorescence spectroscopy (XRF), 139, 43
X-ray photoelectron spectroscopy (XPS), 173, 181
X-ray spectroscopy, see EDS

Yellow metal, 6, 38, 74
Yellow metal corrosion, 38, 74
Yellow metal passivator, 75, 95
Young–Laplace equation, 29